P9-DGB-581

Charles Darwin

THE ORIGIN OF SPECIES

Abridged and Introduced by

PHILIP APPLEMAN

W · W · NORTON & COMPANY · INC · *New York*

Copyright © 1975, 1970 by W. W. Norton & Company, Inc.

FIRST EDITION

Library of Congress Cataloging in Publication Data

Darwin, Charles Robert, 1809–1882.
   The origin of species.

   Bibliography: p.
   Includes index.
   1. Evolution. 2. Natural selection. I. Apple-
man, Philip, 1926-   II. Title.
QH365.02 1975   575.01'62   75–6864
ISBN 0–393–09219–4

PRINTED IN THE UNITED STATES OF AMERICA

1 2 3 4 5 6 7 8 9 0

Contents

The Voyage of the *Beagle* 1831–1836

(The dates of Darwin's landings are not always those of the ship's arrival.)

Plymouth
sailed 27 Dec.31
Falmouth
landed 2 Oct. 36

Off Tenerife
6 Jan. 32
unable to land
because
of quarantine

Azores
landed 20 Sept. 36

St. Paul's Rocks
landed 16 Feb. 32

Cape Verde Is.
landed 16 Jan. 32
landed 31 Aug. 36

Ascension I.
landed 20 July 36

Fernando de Noronha
landed 20 Feb. 32

landed
28 Feb.32

Bahia
landed 1 Aug. 36

St. Helena
landed 8 July 36

Cape Town
landed 1 June 36

Rio de Janeiro
landed 5 Apr. 32

Montevideo
landed 28 July 32

Falkland Is.
landed 1 Mar. 33
landed 10 Mar. 34

Passed out of
Strait of Magellan
10 June 34

Shipped a great sea
13 Jan. 33

Galapagos Is.
landed 16 Sept. 35
sailed 20 Oct. 35

Callao
landed 20 July 35

Valparaiso
landed 23 July 34

Tahiti
landed 15 Nov. 35

Cocos Keeling Is.
landed 2 Apr. 36

Mauritius
landed 30 Apr. 36

King George's Sound
landed 7 Mar. 36

New Zeeland
landed 21 Dec.35

Sydney
landed 12 Jan. 36

Hobart
landed 5 Feb. 36

Introduction

Charles Darwin's youth was unmarked by signs of genius. Born in 1809 into the well-to-do Darwin and Wedgwood clans (his mother was a Wedgwood, and Darwin himself was to marry another), he led a secure and carefree childhood, happy with his family, indifferent to books, responsive to nature. The son and grandson of impressively successful physicians, he eventually tried medical training himself, but found the studies dull and surgery (before anesthesia) too ghastly even to watch. So, for want of anything better, he followed the advice of his awesome father (6'2", 336 pounds, domineering in temperament) and studied for the ministry, taking his B.A. at Christ's College, Cambridge, in 1831.

Then a remarkable turn of events saved Darwin from a country parsonage. His science teacher at Cambridge, John Stevens Henslow, arranged for Darwin the invitation to be naturalist on H.M.S. *Beagle* during a long voyage of exploration. Despite his father's initial reluctance, Darwin got the position, and at the end of 1831 left England for a five-year voyage around the globe that turned out to be not only a crucial experience for Darwin himself, but a passage of consequence for the whole world.

The voyage of the *Beagle* made a scientist of Darwin—an industrious collector, a keen observer, a canny theorist—and it set him a momentous problem that he was to spend the next twenty years struggling with: the problem of the origin of species. By 1858, when another young naturalist, Alfred Russel Wallace, was himself pondering the same problem, Darwin had collected enormous masses of detail relating to species and supporting his theory of natural selection. Faced, after decades of work, with the threat of being anticipated by Wallace's independent discoveries, Darwin quickly finished and published his great work, *The Origin of Species*, in 1859. The effects of that publication were many and profound; hardly any kind of thought—scientific, philosophical, social, literary, historical—remained long unchanged by the implications of the *Origin*.

Except for the voyage of the *Beagle*, however, Darwin's adventures were mostly intellectual, his life deliberately domestic. A chronic sufferer from a mysterious ailment, he had neither the strength nor the temperament for an active and public career; he remained secluded at his country house at Down, shunning the furious post-*Origin* controversies and leaving the defense of "Darwinism" to his more pugnacious friends. But always he worked. Good Victorian that he was, he worked as much every day as his strength permitted, and his industrious life was studded with solid contributions to science in articles, reviews, and books: *The Descent of Man* (1871), *The Expression of the Emotions* (1872), *The Formation of Vegetable Mould* (1881), and so on, and on.

There was something paradoxical but eminently admirable about both Darwin's character and his devotion to his task. Intellectually he was a revolutionary, but the gentlest of revolutionaries. Spiritually he became an agnostic, but never a simple materialist; like many another Victorian agnostic, he exemplified in his life and work a high-minded benevolence not only toward his fellow men but for all creatures, and he continued always to write of the "grandeur" of "beautiful" and "wonderful" forms of life, and of men's high "destiny" in the future.

When he died in 1882, the man whose sacrilegious ideas had once been publicly assailed, from pulpit and periodical, by a multitude of critics, was buried, with the cordial acquiescence of the Dean, in Westminster Abbey—a few feet from the grave of the other immortal among British scientists, Sir Isaac Newton.

II

To be fair to Darwin's originality of mind, we must see him boarding H.M.S. *Beagle* young, amateurish in science, a believer in Genesis. We should picture him carrying in his small library Charles Lyell's new work, the *Principles of Geology*, but warned against its heresies by his respected master, Henslow. We must remember that he carried, too, the lessons of a close study of William Paley's *Natural Theology*. ("I do not think I hardly ever admired a book more than Paley's," he was to write in his autobiography; "I could almost formerly have said it by heart.") And it was Paley more than any other man who had already convinced a whole generation of readers that in the Deity's neatly constructed universe, "the marks of *design* are too strong to be gotten over."

Darwin also carried in his mental baggage the teachings of the distinguished scientist and philosopher William Whewell ("Next to Sir J. Mackintosh . . . the best converser on grave subjects to whom I ever listened"). And by an ironic trick of history, it was during the five years of the *Beagle*'s voyage that the British citadel of scientific respectability, the Royal Society, was administering the publication of the Bridgewater Treatises, a series of books by notable scientists and moralists commissioned specifically to illustrate "the power, wisdom, and goodness of God as manifested in the Creation"—so, while Darwin was examining reptiles and fossils on the coast of Brazil, the grave Dr. Whewell, eminent mathematician and mineralogist, was writing for his Treatise: "If there be, in the administration of the universe, intelligence and benevolence, superintendence and foresight, grounds for love and hope, such qualities may be expected to appear in the constitution and combination of those fundamental regulations by which the course of nature is brought about, and made to be what it is." [1]

It is an awesome distance from that kind of reasoning to the conclusion of the *Origin of Species*: "Thus, from the war of nature, from famine and death, the most exalted object which we are capable of conceiving, namely, the production of the higher animals, directly follows"; and Darwin—cautious, skeptical, compulsively industrious, distrustful of his own talents and never daring to suspect himself of genius—Darwin did not make the voyage in a day.

It was not enough for Darwin that he was pre-eminently an empiricist: that for five years in exotic locales he had explored river beds and coral reefs; hiked pampas and climbed mountains; recorded stratifications of rocks and soil; examined the earth with lens, compass, clinometer, penknife, blowpipe, and acids; discovered fossils of conifers and shellfish, megatherium and mastodon; collected flowers, birds, insects, and reptiles. It was not enough that he was a subtle theorist, pondering the elevation and subsidence of volcanic strata, the causes of the earthquakes he experienced, the formation of coral reefs, and the relation of one species of ground sloth to another. Nor was it even enough that the maturing Darwin, widely respected as a naturalist after the *Beagle* voyage, should have come home to spend two decades in the dogged pursuit of a hypothesis: examining the many breeds of domestic pigeons, the skeletons of rabbits, the wings of ducks, the variations in ten thousand specimens of barnacles; keeping notebooks on "transmutation"; discussing the species problem with close scientific friends; and finally, twenty years after debarking for the last time from the *Beagle*, daring to begin to write his great book on "Natural Selection."

All this was not enough for Darwin, because he understood clearly the strength of conventional scientific opinion on the fixity of species. Lamarck's [2] experience had been an object-lesson: he had

1. William Whewell, *Astronomy and General Physics Considered with Reference to Natural Theology* (London, 1833), pp. 4–5.

2. J. B. de Monet, Chevalier de Lamarck (1744–1829), French naturalist, proposed that environmental conditions produced inheritable changes in the structures of organisms.

challenged this conventional opinion, and his arguments had been systematically attacked by the French scientific establishment and discredited by Lyell. The author of V*estiges of the Natural History of Creation*, that notorious evolutionary publication of 1844, had chosen discretion rather than valor, remaining anonymous. And other scientists, philosophers, and writers (including Darwin's own poetic grandfather) had speculated about the transmutation of species, but their work was never enough, either; it was always too hypothetical, too desultory, too superficial, too limited, too abstract, or too obscure to threaten in any serious way the established Truth of the fixity of species.

So, on June 18, 1858, after five years in the field, twenty years of patient observation and cautious speculation, and two years of busy writing on the manuscript of "Natural Selection," Darwin was still not ready to publish his challenge to prevailing opinion. On that day, however, he received the momentous letter from Alfred Russel Wallace describing Wallace's own recent discovery of the principle of natural selection. Darwin immediately wrote to Lyell: "Your words have come true with a vengeance—that I should be forestalled." But his dismay was temporary: his friends Lyell and Hooker arranged a joint presentation of short papers by both Wallace and Darwin at the Linnean Society in July, 1858, so the names of Darwin and Wallace are permanently linked as co-discoverers of the principle of natural selection. Wallace, however, was always modest about his contribution, since compared to Darwin's it was, as he said, as two weeks are to twenty years.

Darwin then went on, thanks to Wallace's unintentional prompting, and finished an "abstract" of his work: in March, 1859, he completed the *Origin*. It was published on November 24 of that year, and Darwin, tired and sickly, waited for the response.

III

It was a well-known nineteenth-century observation that all great scientific discoveries pass through three stages: in the first stage, people say, "It's absurd"; in the second, "It's contrary to the Bible"; and in the third, "Oh, we've known *that* all along." Evolution passed through all three stages during Darwin's own lifetime, so that in his last revision of the *Origin* he could write:

> As a record of a former state of things, I have retained in the foregoing paragraphs, and elsewhere, several sentences which imply that naturalists believe in the separate creation of species; and I have been much censured for having thus expressed myself. But undoubtedly this was the general belief when the first edition

of the present work appeared. I formerly spoke to very many naturalists on the subject of evolution, and never once met with any sympathetic agreement. It is probable that some did then believe in evolution, but they were either silent, or expressed themselves so ambiguously that it was not easy to understand their meaning. Now things are wholly changed, and almost every naturalist admits the great principle of evolution.[3]

Scientific revolutions depend upon (among other things) the perceptiveness and the industriousness—and sometimes the aggressiveness—of their protagonists. Darwin had had the perception to discover natural selection and the industriousness to fill hundreds of pages with minutely observed facts and close reasoning; friends, upon demand, supplied the aggressiveness; and the world was changed: converted, in a very few years, from an almost total belief in the permanence of species to an almost total belief (among the educated) in the transmutation of species. This in itself was a monumental conversion, but the "Darwinian revolution" implies much more.

It implies, for instance, a transformation of attitudes, of outlook, on the part of other scientists. Since 1859 it has been necessary to think in post-Darwinian terms in order to understand the drift of modern paleontology, comparative anatomy, genetics, ecology, embryology, taxonomy—and in fact virtually all branches of botany and zoology. The "Darwinian revolution" goes well beyond science and scientists, however. It implies a basic change in ways of looking at all ideas, all phenomena. Relativity and flux were not nineteenth-century inventions, of course, but there was an imperiousness about nineteenth-century relativism that was new—new precisely because of the impressiveness of nineteenth-century "science" in general. To the conservative-minded, the terrifying thing about the implications of Darwinism was that nothing was sacrosanct: evolution became not only the science of sciences—worse still, it became the philosophy of philosophies. For those least prepared for change, the impact of all this was shattering. Seventy years after the publication of the *Origin*, Joseph Wood Krutch surveyed the wreckage in his book *The Modern Temper*:

> Three centuries lay between the promulgation of the Copernican theory and the publication of the *Origin of Species*, but in sixty-odd years which have elapsed since that latter event the blows have fallen with a rapidity which left no interval for recovery. The structures which are variously known as mythology, religion, and philosophy, and which are alike in that each has as its function the interpretation of experience in terms which have human values, have collapsed under the force of successive attacks and

3. Charles Darwin, *The Origin of Species* (New York, 1958), p. 445.

shown themselves utterly incapable of assimilating the new stores of experience which have been dumped upon the world. With increasing completeness science maps out the pattern of nature, but the latter has no relation to the pattern of human needs and feelings.[4]

"Mythology, religion, and philosophy" did not succumb without a struggle; counterattacks from the faithful were swift and fierce. Priests, parsons, and bishops defended not only their own faith but also "true Baconian induction," in the pages of the quarterly and theological reviews, where scientific "experts" could flourish in protective anonymity. "The theory of evolution," wrote a truculent Victorian contributor to the *Catholic World*, "has no scientific character, is irreconcilable with the conclusions of natural history, and has no ground to stand upon except the worn-out fallacies of a perverted logic. To call it 'hypothesis' is therefore to do it an honor which it does not deserve. A pile of rubbish is not a palace, and a heap of blunders is not a hypothesis." [5] The aging Anglican sage, Dr. Pusey, sternly repudiated Darwin: "Never probably was any system built upon so many 'perhaps,' 'probably,' 'possibly,' 'it may be,' 'it seems to be,' 'most likely,' 'it must be,' 'it requires but a slight stretch of imagination to conceive,' as that mythological account of the origin of all which has life, and, at last of ourselves, which is now being every where or widely acknowledged by unscientific minds as if it were axiomatic truth; some of whose adherents claim that it will revolutionise every other science." [6] And Pope Pius IX, writing to a French anti-Darwinian author, was thoroughly contemptuous:

> We have received with pleasure, dear son, the work in which you refute so well the aberrations of Darwinism. A system which is repudiated by history, by the traditions of all peoples, by exact science, by the observation of facts, and even by reason itself, would seem to have no need at all of refutation, if alienation from God and the penchant for materialism, both stemming from corruption, were not avidly searching for support in this fabric of fables. . . . But the corruption of this century, the guile of the depraved, the danger of oversimplification demand that such dreamings, absurd as they are, since they wear the mask of science, be refuted by true science. [7]

It need hardly be said that this rather novel concern of theologians for "true science" was not wholly disinterested. Theologians

4. Joseph Wood Krutch, *The Modern Temper* (New York, 1929), p. 12.
5. "Dr. Draper and Evolution," *Catholic World*, XXVI (1878), 775.
6. Edward Bouverie Pusey, *Un-Science, Not Science, Adverse to Faith* (London, 1879), p. 32.
7. Quoted in Constantin James, *L'Hypnotisme expliqué et Mes Entretiens avec S. M. l'Empereur Don Pédro sur le Darwinisme* (Paris, 1888), pp. 84–85; translation by the editor.

worried because they saw, perhaps more clearly than others, the philosophical implications of post-Darwinian thought. It was not just that Darwin had complicated the reading of Genesis; or even that he had furnished impressive scientific authority for the nineteenth-century habit of thinking in terms of wholes and continuities rather than in discrete parts and rigidities; or that the evolutionary orientation stressed context and complexity—though all of these influences could be bothersome when used by "materialists." The worst threat of all was that Darwin's universe operated not by Design but by natural selection, a self-regulating mechanism. ("Paley's divine watchmaker was unemployed," Gavin de Beer writes, "because the wonderful property of organisms is that they make and mar themselves.") Natural selection pictured the world in a constant process of change, but without any apparent prior intention of going anywhere in particular or of becoming anything in particular.

That was a devastating proposition to the conventional theologian—more so, perhaps, than the Copernican theory had been, because it struck so close to home. Natural selection therefore seemed, to many, hopelessly negative, fraught with blasphemy and conducive of despair. Science, with all of its impressiveness of fact and achievement, was moving in on theologians and philosophers, *taking over*: constantly enlarging the domain of fact and consequently reducing the domain of speculation. Given objective knowledge, people tend (it was already clear) to give up the guessing-games of ignorance—conjecture about which of the "humors" is overbalanced in a person whose blood pressure is abnormal; or whether or not Adam and Eve had navels; or what influence the planets have on our destinies. So it *made a difference* to philosophers and theologians that man not only evolved, but evolved by natural selection rather than by a vital force or cosmic urge of some sort. Darwinism seemed uncompromisingly non-teleological, non-vitalist, and non-finalist, and that basic fact could not help but affect the work of philosophers. "Once man was swept into the evolutionary orbit," Bert James Loewenberg has written, "a revolution in Western thought was initiated. Man was seen to be a part of nature, and nature was seen to be a part of man. The Darwinian revolution was not a revolution in science alone; it was a revolution in man's conception of himself and in man's conception of all his works." [8]

With the passing of decades, the tone of the theologians had to be modified and the strategies of philosophers altered: direct opposition to Darwin (in all but the most remote backwaters) gradually

8. Bert James Loewenberg, ed., *Charles Darwin: Evolution and Natural Selection* (Boston, 1959), p. 21.

made way for accommodation. The first edition of the *Catholic Encyclopedia* (1909), for instance, still clung to the older attitude, saying of natural selection that "As a theory, it is scientifically inadequate," and adding, "The third signification of the term *Darwinism* arose from the application of the theory of selection to man, which is likewise impossible of acceptance." [9] By 1967, however, the *New Catholic Encyclopedia* was looking at these things differently: "Today, with a much better understanding of both the theological sources of the Judeo-Christian revelation and of evolutionary theory, the compatibility of God's creative and directive action is more easily comprehended . . . the solution of the basic difficulties [with Darwinism] was soon found to lie in Biblical research and scholarship and not in the rejection of the new theory. . . . In his encyclical *Humani generis* [1950], Pope Pius XII [asserted] that general evolution, even of the body of man (and woman), should be professionally studied by both anthropologist and theologian." [1]

"Present-day theologians," says a recent Catholic commentator, "are far more moderate in their claims than were their predecessors." [2]

The activities of science, relentlessly pushing back the margins of the unknown, have in effect been forcing the concept of "God" into a perpetual retreat into the still-unknown, and it is in this condition that "God" has frequently come to have meaning for modern man. The final retreat is of course into the strongholds of the Infinite and the Eternal, and with the scientists' continuing success in exploring the littleness and vastness of the finite, one may assume that, for any practical purpose, that final retreat has long since occurred. The implications of all this have led some men to a renewed humanism: to the proposition that evolutionary man, cut off from theological presumption, might still have sufficient reason to exist—even to respect himself—simply as man. At the beginning of this century, Bertrand Russell wrote:

> That Man is the product of causes which had no prevision of the end they were achieving; that his origin, his growth, his hopes and fears, his loves and his beliefs, are but the outcome of accidental collocations of atoms; that no fire, no heroism, no intensity of thought and feeling, can preserve an individual life beyond the grave; that all the labours of the ages, all the devotion, all the inspiration, all the noonday brightness of human genius, are destined to extinction in the vast death of the solar system, and that the whole temple of Man's achievement must inevitably be bur-

9. *Catholic Encyclopedia* (New York, 1909), V, 655.
1. *New Catholic Encyclopedia* (New York, 1967), V, 693.
2. Robert W. Gleason. "A Note on Theology and Evolution," in Walter J. Ong, S.J., ed., *Darwin's Vision and Christian Perspectives* (New York, 1960), p. 104.

ied beneath the debris of a universe in ruins—all these things, if not quite beyond dispute, are yet so nearly certain, that no philosophy which rejects them can hope to stand. Only within the scaffolding of these truths, only on the firm foundation of unyielding despair, can the soul's habitation henceforth be safely built.[3]

It is possible to argue that there has been, in consequence of the retreat of "God," a genuine spiritual gain. To discuss "God" in terms of the Unknown seems to some contemporary thinkers (as indeed it did to Herbert Spencer) a more dignified and tenable procedure than to carry on that discussion in terms of oak trees or thunderbolts or man's image. The possible attitudes toward this "Unknown" are various: they can be put in terms of release and new potentiality, as in Bertrand Russell's further comment, "It is well to exalt the dignity of Man, by freeing him as far as possible from the tyranny of non-human Power"; or of guarded optimism, like Julian Huxley's, seeing man as Vicar of Evolution on Earth, a conscious directing factor in his own progress toward an unknown destiny; or of a kind of ecstasy, like Albert Einstein's, in "rapturous amazement" at the harmony of nature. But in any case, men now seem able to view the human condition not in terms of a fall from grace to degradation but rather as the reverse—as a long struggle to escape from mere animalism, red in tooth and claw, and to establish upon man himself—upon his own best knowledge of himself—a tenable ethical ground.

Scientists have rarely been simple materialists, after all. It is important to scientists, as scientists, to restrict their investigations to the knowable; but they have not often contended that the results of their researches were the summation of all Truth, or that one should act as though the world were purely deterministic. Moreover, some scientists, like T. H. Huxley and his grandson, Julian Huxley, have been consistently interested in the relationship between science (particularly evolution) and ethics. Their conclusions have been hopeful, not pessimistic, and one may cite, as illustrative examples, both T. H. Huxley's belief that "there lies within [man] a fund of energy, operating intelligently and so far akin to that which pervades the universe, that it is competent to influence and modify the cosmic process" and Julian Huxley's insistence that "in so far as the mechanism of evolution ceases to be blind and automatic and becomes conscious, ethics can be injected into the evolutionary process. Before man that process was merely amoral. After his emergence onto life's stage it became possible to introduce faith, courage, love of truth, goodness—in a word moral purpose

3. Bertrand Russell, *Mysticism and Logic* (New York, 1957), pp. 45–46.

—into evolution." [4] So it is not after all strange, or even unusual, to hear the very scientists who remind man of his cosmic littleness at the same time assuring him of his individual dignity in terms which, if not so dogmatic as those of the theologians, are nevertheless heartening.

IV

As some of the foregoing suggests, the moment one transplants the discussion of evolution from the field of science to the fields of philosophy or ethics or theology, one has to recognize that the discussion has probably become analogical rather than straightforwardly analytical, "Darwinistic" rather than Darwinian. ("Darwinism," writes Morse Peckham, "is a scientific theory about the origin of biological species from pre-existent species. . . . Darwinisticism can be an evolutionary metaphysic about the nature of reality and the universe. It can be an economic theory, or a moral theory, or an aesthetic theory, or a psychological theory." [5]) Darwin's propositions, like many other bright and fashionable ideas before and since, caught people's imaginations; and many thinkers picked up the new concepts and exploited them as analogies (or apologies) in their own fields of work, often less in the hope of being led to new perceptions by the analogies than with the intention of reinforcing old positions or of scoring debater's points with them. Whitehead reflected on the dangers of this indiscriminate and tendentious use of Darwinistic ideas. "The last words of science," he wrote, "appeared to be Struggle for Existence and Natural Selection. Darwin's own writings are for all time a model of refusal to go beyond the direct evidence, and of careful retention of every hypothesis. But those virtues were not so conspicuous in his followers, and still less in his camp-followers." [6]

We see this tendency perhaps most clearly in the application of a Darwinistic pattern to society itself. The publication of the *Origin* and the American Civil War were almost coincident, and in the years following those two traumatic events, the United States was industrializing very rapidly. The late nineteenth century became the pre-eminent period in our history of the Rugged Individualist, the Robber Baron, the Captain of Industry, the accumulators of great wealth. It was also a period of sweatshops, of union-busting, of goon squads and strike-breaking massacres, of a dollar a day for workingmen, of tenements without sanitation, of widespread malnutrition.

4. Julian Huxley and T. H. Huxley, *Touchstone for Ethics* (New York and London, 1947), p. 135.
5. Morse Peckham, "Darwinism and Darwinisticism," *Victorian Studies*, III (1959), 32.
6. Alfred North Whitehead, *Science and the Modern World* (New York, 1925), p. 158.

It was not simple, in a "Christian society," to reconcile such contradictions. The economic establishment had, since the beginnings of the industrial revolution, been casting about for sanctions, for justifications. Manchester political economy had been a powerful sanction for the establishment, emphasizing as it did the necessity for untrammelled individual enterprise, the automatic enlightenment of self-interest, and the "iron" laws of economics. ("There is no more possibility of defeating the operation of these laws," Andrew Carnegie once said, "than there is of thwarting the laws of nature which determine the humidity of the atmosphere or the revolution of the earth upon its axis." [7]) Curiously enough, there were religious sanctions, too, for had not material things always been corrupting to man, and was it not therefore self-evident that the lower classes had to be kept poor to be kept virtuous—and, paradoxically, could not the virtuous and industrious among the poor expect to be rewarded, even in this world? ("He that gets all he can honestly," Ben Franklin had declared, "and saves all he gets will certainly become rich, if that Being who governs the world, to whom all should look for a blessing on their endeavors, doth not, in His wise providence, otherwise determine." To which a railroad president added, a century later: "The rights and interests of the laboring man will be protected and cared for by the Christian men to whom God has given control of the property rights of the country." [8]

And always, in the nineteenth century, there was that court of last resort, the sanction of Progress, in whose name all contradictions were resolved—or ignored.

But in natural selection the economic establishment found a massive sanction which gathered into one grand synthesis all of the previous ones and added to them its own profound, "scientific" prestige. Natural selection and the struggle for existence lent authority to laissez-faire economics (the state, said Herbert Spencer, should refrain from action calculated to interfere with the struggle for existence in the industrial field [9]); absorbed the older religious sanctions (the "laws" of natural selection, said one economist, were "merely God's regular methods of expressing his choice and approval" [1]); and seemed to make Progress inevitable (evolution, said Spencer, "can end only in the establishment of the greatest perfection and the most complete happiness"; "Progress," he said, "is not an accident, but a necessity" [2]).

7. Andrew Carnegie, *The Empire of Business* (London, 1902), p. 67.
8. Quoted in Matthew Josephson, *The Robber Barons* (New York, 1934), p. 299.
9. Edward S. Corwin, "The Impact of the Idea of Evolution on the American Political and Constitutional Tradition," in Stow Persons, ed., *Evolutionary Thought in America* (New Haven, 1950), p. 186.
1. Thomas Nixon Carver, quoted in Richard Hofstadter, *Social Darwinism in American Thought* (Boston, 1955), p. 40.
2. Hofstadter, p. 40.

Among American economists, the most thoroughgoing advocate of these views was William Graham Sumner, whose reading of Spencer in the early 1870's convinced him that Spencer was right in seeing the world as "a harsh, exacting nature, enforcing a bitter struggle for the meager goods available to humankind. To him the laws of individual and social existence were simple and rigorous. Rewards and punishments are meted out with impartial justice. Property, the enjoyment of family life, health, social preference go to the fit, while poverty, disease, and starvation are the lot of the unfit." And thus "the millionaires are a product of natural selection, acting on the whole body of men to pick out those who can meet the requirement of certain work to be done." [3] This is conservative "social Darwinism" in a nutshell. (I put the phrase in quotation marks to indicate its spuriousness: Darwin should not, of course, be held responsible for this—or any other—brand of Darwinisticism.)

The Captains of Industry were quick to pick up this Darwinistic vocabulary. "The growth of a large business," said John D. Rockefeller (speaking to a Sunday-school class), "is merely a survival of the fittest. . . . The American Beauty rose can be produced in the splendor and fragrance which bring cheer to the beholder only by sacrificing the early buds which grow up around it. This is not an evil tendency in business. It is merely the working-out of a law of nature and of God." [4] Similarly, James J. Hill held that "the fortunes of railroad companies are determined by the law of the survival of the fittest." [5] And Carnegie once warned his lieutenants, upon his break with the Steel Pool, that "a struggle is inevitable and it is a question of the survival of the fittest." [6]

Carnegie was the most introspective and easily the most articulate of the Rugged Individualists. He was strongly attracted to Spencer's theories, and his own speech and writings became full of Spencer's phraseology. He quickly adopted Spencer's easy optimism about progress; he believed firmly that the "fit" who survived in the struggle were the "best," and that these men would lead the rest to certain progress. "The exceptional man in every department," he said, "must be permitted and encouraged to develop his unusual powers, tastes, and ambitions in accordance with the laws which prevail in everything that lives or grows. The 'survival of the fittest' means that the exceptional plants, animals, and men which have the needed 'variations' from the common standard, are the fructifying forces which leaven the whole." [7]

3. See Maurice R. Davis, ed., *Sumner Today* (New Haven, 1940), p. 92, and Hofstadter, p. 58.
4. Quoted in Robert E. L. Faris, "Evolution and American Sociology," in Persons, p. 163.
5. James J. Hill, quoted in Hofstadter, p. 45.
6. Josephson, p. 420.
7. Andrew Carnegie, *Problems of Today* (New York, 1933), p. 125.

But there were all sorts of anomalies in this use of the evolutionary sanctions: anomalies of language, for instance. To scientists, the word "natural" meant simply "the way things occur in nature"; but used by the apologists for "social Darwinism," it always implied "the way things ought to be." "Survival" was similarly converted into a term of generalized approbation—so persistently that William James was prompted to object: "The entire modern deification of survival *per se*, survival returning to itself, survival naked and abstract, with the denial of any substantive excellence in *what* survives, except the capacity for more survival still, is surely the strangest intellectual stopping-place ever proposed by one man to another." [8] The most abused of all of the evolutionary terms, however, was the word "fit." Spencer and Sumner insisted that the wealthy must be the "fit," for obvious reasons, and conversely, that the poor must be the "unfit," since the latter were, as Spencer held, the "stupid, vicious, or idle," or, to use Sumner's words, the "negligent, shiftless, inefficient, silly, and imprudent." The fallacies of this terminology are obvious and have often been objected to. T. H. Huxley insisted that the biologically "fittest" were by no means always the "highest" order of creatures, and G. G. Simpson has pointed out that, inchoately in Darwin and definitely for neo-Darwinians, "fittest" means simply: leaving the most descendants over a number of generations. Lester Ward, from a less biological, more moralistic point of view, observed that Sumner's definition of the "fit" was based on the "fundamental error that the favors of this world are distributed entirely according to merit." [9]

Another anomaly in the way the conservative "social Darwinists" used Darwinistic analogies was in their tendency to apply them selectively. Capitalistic theory required a "struggle for existence" among competing economic organizations; in practice, however, those very Rugged Individualists who appealed to the competitive Darwinistic analogy against social reforms often tried to thwart its effects upon themselves. They secured the passage of tariffs to protect their individualism from foreign competition, for instance, and tried hard to eliminate all domestic production struggles. Rockefeller, for one, seemed "convinced that the competitive system . . . was a mistake," that "it was a crime against order, efficiency, economy"; and he came to the characteristic conclusion that "it could be eliminated only by abolishing all rivals." [1] Even Carnegie, "individualist" that he proclaimed himself, nevertheless operated a working monopoly in steel. Walter Lippmann has commented wryly that "most men have shown in their behavior that they

---

8. Quoted in Hofstadter, p. 201.
9. Quoted in Hofstadter, p. 79.

1. John T. Flynn, *God's Gold* (New York, 1932), p. 221.

wished to impose free capitalism on others and to escape it themselves." [2]

It stands as at least a qualified tribute to the humanity of human beings that many people simply would not accept the callousness of conservative "social Darwinism," however much it attempted to justify itself by the appeal to "scientific" prestige. Richard Tawney once argued that one sees the true character of a social philosophy most clearly in "the way it regards the misfortunes of those of its members who fall by the way." [3] Those who fell by the way in late nineteenth-century America were assured by the conservative apologists that this was their "natural" lot, that "science" proved that they were the unavoidable by-products of a beneficent Struggle for Existence, without which there could be no Progress; but meanwhile, men with less emotional investment in maintaining the economic status quo were questioning whether the implications of evolution were really as somberly deterministic as the conservatives made out. As early as the 1870's some zoologists had been investigating the implications of cooperation, as well as competition, in the natural world, and in 1902 Peter Kropotkin was to publish his *Mutual Aid*, revealing that he had

> failed to find—although I was eagerly looking for it—that bitter struggle for the means of existence, *among animals belonging to the same species*, which was considered by most Darwinists (though not always by Darwin himself) as the dominant characteristic of struggle for life, and the main factor of evolution. . . . On the other hand, wherever I saw animal life in abundance . . . I saw Mutual Aid and Mutual Support carried on to an extent which made me suspect in it a feature of the greatest importance for the maintenance of life, the preservation of each species, and its further evolution. [4]

And he generalized, from this pattern, that

> it is not love and not even sympathy upon which Society is based in mankind. It is the conscience—be it only at the stage of an instinct—of human solidarity. It is the unconscious recognition of the force that is borrowed by each man from the practice of mutual aid; of the close dependency of every one's happiness upon the happiness of all; and of the sense of justice, or equity, which brings the individual to consider the rights of every other individual as equal to his own. Upon this broad and necessary foundation the still higher moral feelings are developed. [5]

Thomas Henry Huxley had already made the famous distinction,

2. Walter Lippmann, *The Method of Freedom* (New York, 1934), p. 25.
3. Richard Tawney, *Religion and the Rise of Capitalism* (New York, 1926), p. 247.
4. Peter Kropotkin, *Mutual Aid* (London, 1902), pp. vii–ix.
5. Kropotkin, pp. xiii–xiv.

in his Romanes lecture of 1893, that "the ethical progress of society depends, not on imitating the cosmic process, still less in running away from it, but in combating it." [6] "Social progress," he said, "means a checking of the cosmic process at every step and the substitution for it of another, which may be called the ethical process; the end of which is not the survival of those who may happen to be the fittest, in respect of the whole of the conditions which obtain, but of those who are ethically the best." [7]

By the end of the century, then, there were developing some serious demurrers to narrowly conservative Darwinistic social thought. Some naturalists were broadening the evolutionary view of life to show that nature was not always and simply competitive. And some social and ethical thinkers were attempting to demonstrate that even though ruthless competition does exist in nature, it is not therefore necessarily a proper pattern for human behavior.

V

Discussions of Darwinism and Darwinisticism inevitably return to the *impressiveness* of science, its ineluctable ability to demonstrate, to convince, its power to change the conditions and directions of our lives. Literary men, heirs to thousands of years of a general cultural hegemony, have not always welcomed the intrusion of this new power into their lives. The Victorian author George Gissing once wrote:

> I hate and fear "science" because of my conviction that for long to come if not forever, it will be the remorseless enemy of mankind. I see it destroying all simplicity and gentleness of life, all beauty of the world; I see it restoring barbarism under the mask of civilization; I see it darkening men's minds and hardening their hearts; I see it bringing a time of vast conflicts, which will pale into insignificance "the thousand wars of old," and, as likely as not, will whelm all the laborious advances of mankind in blood-drenched chaos. [8]

Writers' responses to Darwin in particular were often hostile. In novel, story, and poem, frontal attacks and attacks by innuendo and ridicule appeared regularly throughout the last decades of the nineteenth century. But thoughtful writers were at the same time busy assimilating the evolutionary doctrine into their work. Some had been using evolution before 1859: Tennyson was proud of having discussed evolution in his verse not only before the *Origin of Species* but even before the *Vestiges of Creation*. Browning, too, had mentioned evolutionary ideas prior to 1859, and both poets

6. *Touchstone for Ethics*, p. 92.
7. *Touchstone for Ethics*, p. 91.

8. George Gissing, *The Private Papers of Henry Ryecroft*, Chapter 18.

continued to write about the subject, off and on, for the rest of their lives. After 1859, however, evolution cropped up everywhere: in the novels of Bulwer-Lytton, Charles Reade, Mrs. Humphry Ward, Gissing, and Charles Kingsley and in scores of satires and romances, evolution was cited, discussed, and absorbed.

The most distinguished propagators of evolution to the late-Victorian reading public were George Meredith, Thomas Hardy, and A. C. Swinburne, and (in a rather different way and somewhat later) Samuel Butler, H. G. Wells, and George Bernard Shaw. The first three (despite Hardy's gloom) tended to see the outcome of man's tribulations, in a world that is red in tooth and claw, as true progress. Meredith built a "philosophy of earth" around this belief, even though at times he, like Tennyson, only faintly trusted the larger hope that somehow, eventually, good would be the final goal of ill. Equally familiar is the story of how Butler, Wells, and Shaw all turned consciously away from Darwin and toward a Neo-Lamarckian version of evolution. The Lamarckian thesis was more attractive to them than Darwin's natural selection because it seemed more "optimistic"—it allowed more easily a teleological interpretation, a hope that somehow a Purpose did guide our steps aright.

Some writers needed this kind of reassurance; others did not. The influence of Darwinistic thinking upon late-nineteenth-century naturalism is too well known to need recounting here, but it might be recorded that the naturalists saw themselves as tough-minded students of the nature of evolutionary man, observers of his environmental influences, of the objective facts of life. Darwin had revealed man's true corporeal history, and the naturalistic writers pursued this record of the flesh with the intensity of devoted researchers. "Given a strong man and an unsatisfied woman," Zola proposed to "seek in them the beast, to see nothing but the beast . . ."

Explicit adaptations of evolutionary notions into the practice and study of the arts were characteristic of the late nineteenth century, but have been far less so of the twentieth. In part at least this is because the nineteenth century had to go through the difficult process of changing its mind, of adapting itself to evolution; the twentieth century simply accepts it. Evolution has become one of those topics so well founded as not really to be exciting to educated people any more, and consequently direct references to evolution in literature are almost as rare as direct references to the laws of gravity. The indirect influence of evolution on our literature has, on the other hand, been tremendous. Whitehead once proposed, in a much-repeated statement, that students of the history of ideas should not look for those ideas which are under constant discussion in any age, but instead should look for those basic assumptions which are so fundamental to a man's way of thinking that he does not even realize he is assuming them. Evolution has by now become such an unconscious assumption in our society.

One of the supposed literary effects of Darwinian assumptions has already been mentioned: Joseph Wood Krutch (among others) has maintained that Darwin's ideas have eroded human values and therefore destroyed high tragedy in our century. "With increasing completeness," Krutch wrote in *The Modern Temper*, "science maps out the pattern of nature, but the latter has no relation to the pattern of human needs and feelings." This fact, Krutch said, precludes tragedy because "the Tragic Fallacy depends ultimately upon the assumption which man so readily makes that something outside his own being . . . confirms him in his feeling that his passions and his opinions are important. . . . We can no longer tell tales of noble men because we do not believe that noble men exist." [9]

Looking about us from the perspective of the last third of the twentieth century, we can recognize the symptoms of Krutch's diagnosis. Nevertheless, the case as he states it can hardly be the whole story. For one thing, we ought to consider the negative instances. If Darwin destroyed twentieth-century tragedy, who then was responsible for the palpable failure of early nineteenth-century tragedy, or for the failure of eighteenth-century tragedy? In fact, the occurrence of high tragedy has always been far rarer than its non-occurrence; to expect every age to produce a Sophocles or a Shakespeare is overoptimistic.

For another thing: it is not simply axiomatic that a physically mechanistic world is necessarily a morally valueless one. Those who propose that it is seem to regret that natural selection is an automatic process, that it acts independently of teleological considerations, and that it, together with Mendelian genetics, makes such a satisfactory explanation of organic evolution; and they sometimes attempt to ignore all this and postulate instead some sort of "will" or "preference" or "emergent evolution." Since Darwin, however, reasonable men have simply had to abandon naturalistic teleology—one cannot, any more, look to nature for revealing patterns of goal-directed process. Or, to put it another way, goal-direction in nature must now be divorced from final causes or preternatural factors. However, as John Herman Randall, Jr., has pointed out,

> nature is once more for us, as for the Greeks, full of *implicit* ends and ideals, full of "values," just because it is now an affair of processes, of means effecting ends, of things that are "necessary for," "better" and "worse for" other things. It contains so much

9. Krutch, pp. 12, 136–137.

"natural teleology," in terms of which its various factors can be "evaluated." It takes but a single flower to refute the contention that there are no "values" in nature, no achievement of ends through valuable means. We may even say it is obviously "good for" the planet to go round the sun. Of course, neither the flower nor the planet "finds" it good: only men "find" anything. But surely it does not follow that because only men find anything good or bad, better or worse, what they find is not found. The finding is a genuine cooperation with nature.

It is such a nature our best post-Darwinian knowledge and thought now reveal to us.[1]

I would propose not that Darwinism has killed human values and therefore great and tragic art in our time, but something nearer the reverse: that since Darwin, we have been forced, in art as in life, to mature to finiteness. I mean maturing as human beings not simply by realizing that there are no usable absolutes for man—for that mere realization is still a kind of adolescence—but by accepting one's finiteness and learning to live with it with some degree of sanity and integrity.

Oedipus and Lear are not great tragic heroes simply because they stand in relation to the gods as flies to wanton boys, after all, but because of their own impressive human characters. And man has not stopped defending his uniqueness. If he is unique because of a certain type of brain rather than some supernatural prerogative, and if he has chosen a perhaps disagreeable scientific fact in preference to a pretty illusion, the case is nevertheless the same: man's uniqueness, his dignity, remain important to him. Twentieth-century literature taken as a whole—despite absurdity, disillusion, black humor, anti-heroism, existential angst, and cosmic nausea—remains a human-centered, human-valued literature. We need not look for the tragic spirit only in drama. The greatest of the modern writers in all genres—Lawrence, Joyce, Faulkner, Camus, Hemingway, Yeats, Frost, Pirandello, Brecht, Beckett—have kept aspiring man as at least the implied focus of their work, so that even when he reads as a gray failure, we must take him to be the ghostly negative of an implicitly positive picture.

Looking back at a million years of man's struggle to be human, at his errant and painful attempts to be a special kind of animal—the animal who thinks, the animal who creates—it seems to me that despite his shortcomings, we must remain man's admirer. "Thought makes the whole dignity of man," Pascal remarked; and

1. John Herman Randall, Jr., "The Changing Impact of Darwin on Philos-ophy," *Journal of the History of Ideas*, XXII (1961), 458–459.

despite our sad reflections on the inadequacy of our thinking, on its stunted and twisted travesties in history and our daily lives, we must end by agreeing with Pascal. *Homo* is unique and valuable precisely because he is *sapiens*. He is worth keeping because, given his remarkable past, we may continue to hope that he has, as Darwin surmised, "a still higher destiny in the future."

CHARLES DARWIN

The Origin of Species (1859) †

Darwin's Introduction

When on board H.M.S. 'Beagle,' as naturalist, I was much struck with certain facts in the distribution of the organic beings inhabiting South America, and in the geological relations of the present to the past inhabitants of that continent. These facts, as will be seen in the latter chapters of this volume, seemed to throw some light on the origin of species—that mystery of mysteries, as it has been called by one of our greatest philosophers. On my return home, it occurred to me, in 1837, that something might perhaps be made out on this question by patiently accumulating and reflecting on all sorts of facts which could possibly have any bearing on it. After five years' work I allowed myself to speculate on the subject, and drew up some short notes; these I enlarged in 1844 into a sketch of the conclusions, which then seemed to me probable: from that period to the present day I have steadily pursued the same object. I hope that I may be excused for entering on these personal details, as I give them to show that I have not been hasty in coming to a decision.

My work is now (1859) nearly finished; but as it will take me many more years to complete it, and as my health is far from strong, I have been urged to publish this Abstract. I have more especially been induced to do this, as Mr. Wallace, who is now studying the natural history of the Malay archipelago, has arrived at almost exactly the same general conclusions that I have on the origin of species. In 1858 he sent me a memoir on this subject, with a request that I would forward it to Sir Charles Lyell, who sent it to the Linnean Society, and it is published in the third volume of the Journal of that society. Sir C. Lyell and Dr. Hooker, who both knew of my work—the latter having read my sketch of 1844—honoured me by thinking it advisable to publish, with Mr. Wallace's excellent memoir, some brief extracts from my manuscripts.

This Abstract, which I now publish, must necessarily be imperfect. I cannot here give references and authorities for my several statements; and I must trust to the reader reposing some confidence in my accuracy. No doubt errors will have crept in, though I hope I have always been cautious in trusting to good authorities alone. I can here give only the general conclusions at which I have arrived, with a few facts in illustration, but which, I hope, in most cases

† The present text is excerpted from the sixth edition of the *Origin* (1872); the last edition during Darwin's lifetime.

will suffice. No one can feel more sensible than I do of the necessity of hereafter publishing in detail all the facts, with references, on which my conclusions have been grounded; and I hope in a future work to do this. For I am well aware that scarcely a single point is discussed in this volume on which facts cannot be adduced, often apparently leading to conclusions directly opposite to those at which I have arrived. A fair result can be obtained only by fully stating and balancing the facts and arguments on both sides of each question; and this is here impossible.

I much regret that want of space prevents my having the satisfaction of acknowledging the generous assistance which I have received from very many naturalists, some of them personally unknown to me. I cannot, however, let this opportunity pass without expressing my deep obligations to Dr. Hooker, who, for the last fifteen years, has aided me in every possible way by his large stores of knowledge and his excellent judgment.

In considering the Origin of Species, it is quite conceivable that a naturalist, reflecting on the mutual affinities of organic beings, on their embryological relations, their geographical distribution, geological succession, and other such facts, might come to the conclusion that species had not been independently created, but had descended, like varieties, from other species. Nevertheless, such a conclusion, even if well founded, would be unsatisfactory, until it could be shown how the innumerable species inhabiting this world have been modified, so as to acquire that perfection of structure and coadaptation which justly excites our admiration. Naturalists continually refer to external conditions, such as climate, food, &c., as the only possible source of variation. In one limited sense, as we shall hereafter see, this may be true; but it is preposterous to attribute to mere external conditions, the structure, for instance, of the woodpecker, with its feet, tail, beak and tongue, so admirably adapted to catch insects under the bark of trees. In the case of the mistletoe, which draws its nourishment from certain trees, which has seeds that must be transported by certain birds, and which has flowers with separate sexes absolutely requiring the agency of certain insects to bring pollen from one flower to the other, it is equally preposterous to account for the structure of this parasite, with its relations to several distinct organic beings, by the effects of external conditions, or of habit, or the volition of the plant itself.

It is, therefore, of the highest importance to gain a clear insight into the means of modification and coadaptation. At the commencement of my observations it seemed to me probable that a careful study of domesticated animals and of cultivated plants would offer the best chance of making out this obscure problem. Nor have I been disappointed; in this and in all other perplexing cases I have invariably found that our knowledge, imperfect though it be, of variation under domestication, afforded the best and safest clue. I may venture to express my conviction of the high value of such studies, although they have been very commonly neglected by naturalists.

From these considerations, I shall devote the first chapter of this Abstract to Variation under Domestication. We shall thus see that a large amount of hereditary modification is at least possible; and, what is equally or more important, we shall see how great is the power of man in accumulating by his Selection successive slight variations. I will then pass on to the variability of species in a state of nature; but I shall, unfortunately, be compelled to treat this subject far too briefly, as it can be treated properly only by giving long catalogues of facts. We shall, however, be enabled to discuss what circumstances are most favourable to variation. In the next chapter the Struggle for Existence amongst all organic beings throughout the world, which inevitably follows from the high geometrical ratio of their increase, will be considered. This is the doctrine of Malthus, applied to the whole animal and vegetable kingdoms. As many more individuals of each species are born than can possibly survive; and as, consequently, there is a frequently recurring struggle for existence, it follows that any being, if it vary however slightly in any manner profitable to itself, under the complex and sometimes varying conditions of life, will have a better chance of surviving, and thus be *naturally selected*. From the strong principle of inheritance, any selected variety will tend to propagate its new and modified form.

This fundamental subject of Natural Selection will be treated at some length in the fourth chapter; and we shall then see how Natural Selection almost inevitably causes much Extinction of the less improved forms of life, and leads to what I have called Divergence of Character. In the next chapter I shall discuss the complex and little known laws of variation. In the five succeeding chapters, the most apparent and gravest difficulties in accepting the theory will be given: namely, first, the difficulties of transitions, or how a simple being or a simple organ can be changed and perfected into a highly developed being or into an elaborately constructed organ; secondly, the subject of Instinct, or the mental powers of animals; thirdly, Hybridism, or the infertility of species and the fertility of varieties when intercrossed; and fourthly, the imperfection of the Geological Record. In the next chapter I shall consider the geological succession of organic beings throughout time; in the twelfth and thirteenth, their geographical distribution throughout space; in the fourteenth, their classification or mutual affinities, both when mature and in an embryonic condition. In the last chapter I shall give a brief recapitulation of the whole work, and a few concluding remarks.

No one ought to feel surprise at much remaining as yet unexplained in regard to the origin of species and varieties, if he make due allowance for our profound ignorance in regard to the mutual relations of the many beings which live around us. Who can explain why one species ranges widely and is very numerous, and why another allied species has a narrow range and is rare? Yet these relations are of the highest importance, for they determine the present welfare and, as I believe, the future success and modification of every inhabitant of this world. Still less do we know of the mutual relations of the innumerable inhabitants of the world during the many past geological epochs in its history. Although much remains obscure, and will long remain obscure, I can entertain no doubt, after the most deliberate study and dispassionate judgment of which I am capable, that the view which most naturalists until recently entertained, and which I formerly entertained—namely, that each species has been independently created—is erroneous. I am fully convinced that species are not immutable; but that those belonging to what are called the same genera are lineal descendants of some other and generally extinct species, in the same manner as the acknowledged varieties of any one species are the descendants of that species. Furthermore, I am convinced that Natural Selection has been the most important, but not the exclusive, means of modification.

Chapter I

VARIATION UNDER DOMESTICATION

Causes of variability —Effects of habit and the use or disuse of parts—Correlated variation—Inheritance—Character of domestic varieties—Difficulty of distinguishing between varieties and species —Origin of domestic varieties from one or more species—Domestic pigeons, their differences and origin—Principles of selection, anciently followed, their effects—Methodical and unconscious selection—Unknown origin of our domestic productions—Circumstances favourable to man's power of selection.

CAUSES OF VARIABILITY

When we compare the individuals of the same variety or subvariety of our older cultivated plants and animals, one of the first points which strikes us is, that they generally differ more from each other than do the individuals of any one species or variety in a state of nature. And if we reflect on the vast diversity of the plants and animals which have been cultivated, and which have varied during all ages under the most different climates and treatment, we are driven to conclude that this great variability is due to our domestic productions having been raised under conditions of life not so uniform as, and somewhat different from, those to which the parent species had been exposed under nature. There is, also, some probability in the view propounded by Andrew Knight, that this variability may be partly connected with excess of food. It seems clear that organic beings must be exposed during several generations to new conditions to cause any great amount of variation; and that, when the organisation has once begun to vary, it generally continues varying for many generations. No case is on record of a variable organism ceasing to vary under cultivation. Our oldest cultivated plants, such as wheat, still yield new varieties: our oldest domesticated animals are still capable of rapid improvement or modification.

As far as I am able to judge, after long attending to the subject, the conditions of life appear to act in two ways,—directly on the whole organisation or on certain parts alone, and indirectly by affecting the reproductive system. With respect to the direct action, we must bear in mind that in every case, as Professor Weismann has lately insisted, and as I have incidentally shown in my work on 'Variation under Domestication,' there are two factors: namely, the nature of the organism, and the nature of the conditions. The former seems to be much the more important; for nearly similar variations sometimes arise under, as far as we can judge, dissimilar conditions; and, on the other hand, dissimilar variations arise under conditions which appear to be nearly uniform. The effects on the offspring are either definite or indefinite. They may be considered as definite when all or nearly all the offspring of individuals exposed to certain conditions during several generations are modified in the same manner. It is extremely difficult to come to any conclusion in regard to the extent of the changes which have been thus definitely induced. There can, however, be little doubt about many slight changes,—such as size from the amount of food, colour from the nature of the food, thickness of the skin and hair from climate, &c. Each of the endless variations which we see in the plumage of our fowls must have had some efficient cause; and if the same cause were to act uniformly during a long series of generations on many individuals, all probably would be modified in the same manner. * * *

EFFECTS OF HABIT AND OF THE USE OR DISUSE OF PARTS; CORRELATED
VARIATION; INHERITANCE

* * * The laws governing inheritance are for the most part unknown. No one can say why the same peculiarity in different individuals of the same species, or in different species, is sometimes inherited and sometimes not so; why the child often reverts in certain characters to its grandfather or grandmother or more remote ancestor; why a peculiarity is often transmitted from one sex to both sexes, or to one sex alone, more commonly but not exclusively to the like sex. It is a fact of some importance to us, that peculiarities appearing in the males of our domestic breeds are often transmitted, either exclusively or in a much greater degree, to the males

alone. A much more important rule, which I think may be trusted, is that, at whatever period of life a peculiarity first appears, it tends to reappear in the offspring at a corresponding age, though sometimes earlier. In many cases this could not be otherwise; thus the inherited peculiarities in the horns of cattle could appear only in the offspring when nearly mature; peculiarities in the silkworm are known to appear at the corresponding caterpillar or cocoon stage. But hereditary diseases and some other facts make me believe that the rule has a wider extension, and that, when there is no apparent reason why a peculiarity should appear at any particular age, yet that it does tend to appear in the offspring at the same period at which it first appeared in the parent. I believe this rule to be of the highest importance in explaining the laws of embryology. * * *

CHARACTER OF DOMESTIC VARIETIES; DIFFICULTY OF DISTINGUISHING BETWEEN VARIETIES AND SPECIES; ORIGIN OF DOMESTIC VARIETIES FROM ONE OR MORE SPECIES

When we look to the hereditary varieties or races of our domestic animals and plants, and compare them with closely allied species, we generally perceive in each domestic race, as already remarked, less uniformity of character than in true species. Domestic races often have a somewhat monstrous character; by which I mean, that, although differing from each other, and from other species of the same genus, in several trifling respects, they often differ in an extreme degree in some one part, both when compared one with another, and more especially when compared with the species under nature to which they are nearest allied. With these exceptions (and with that of the perfect fertility of varieties when crossed,—a subject hereafter to be discussed), domestic races of the same species differ from each other in the same manner as do the closely-allied species of the same genus in a state of nature, but the differences in most cases are less in degree. This must be admitted as true, for the domestic races of many animals and plants have been ranked by some competent judges as the descendants of ab- originally distinct species, and by other competent judges as mere varieties. If any well marked distinction existed between a domestic race and a species, this source of doubt would not so perpetually recur. It has often been stated that domestic races do not differ from each other in character of generic value. It can be shown that this statement is not correct; but naturalists differ much in de- termining what characters are of generic value; all such valuations being at present empirical. When it is explained how genera orig- inate under nature, it will be seen that we have no right to expect often to find a generic amount of difference in our domesticated races.

In attempting to estimate the amount of structural difference between allied domestic races, we are soon involved in doubt, from not knowing whether they are descended from one or several parent species. This point, if it could be cleared up, would be interesting; if, for instance, it could be shown that the greyhound, bloodhound, terrier, spaniel, and bull-dog, which we all know propagate their kind truly, were the offspring of any single species, then such facts would have great weight in making us doubt about the immutability of the many closely allied natural species—for instance, of the many foxes—inhabiting different quarters of the world. I do not believe, as we shall presently see, that the whole amount of difference between the several breeds of the dog has been produced under domestication, I believe that a small part of the difference is due to their being descended from distinct species. In the case of strongly marked races of some other domesticated species, there is presumptive or even strong evidence, that all are descended from a single wild stock. * * *

BREEDS OF THE DOMESTIC PIGEON, THEIR DIFFERENCES AND ORIGIN

Believing that it is always best to study some special group, I have, after deliberation, taken up domestic pigeons. I have kept every breed which I could purchase or obtain, and have been most kindly favoured with skins from several quarters of the world, more especially by the Hon. W. Elliot from India, and by the Hon. C. Murray from Persia. Many treatises in different languages have been published on pigeons, and some of them are very important, as being of considerable antiquity. I have associated with several eminent fanciers, and have been permitted to join two of the London Pigeon Clubs. The diversity of the breeds is something astonishing. Compare the English carrier and the short-faced tumbler, and see the wonderful difference in their beaks, entailing corresponding differences in their skulls. The carrier, more especially the male bird, is also remarkable from the wonderful development of the carunculated skin about the head; and this is accompanied by greatly elongated eyelids, very large external orifices to the nostrils, and a wide gape of mouth. The short-faced tumbler has a beak in outline almost like that of a finch; and the common tumbler has the singular inherited habit of flying at a great height in a compact flock, and tumbling in the air head over heels. The runt is a bird of great size, with long massive beak and large feet; some of the sub-breeds of runts have very long necks, others very long wings and tails, others singularly short tails. The barb is allied to the carrier, but, instead of a long beak has a very short and broad one. The pouter has a much elongated body, wings, and legs; and its enor-

mously developed crop, which it glories in inflating, may well
excite astonishment and even laughter. The turbit has a short and
conical beak, with a line of reversed feathers down the breast; and
it has the habit of continually expanding slightly, the upper part of
the œsophagus. The Jacobin has the feathers so much reversed
along the back of the neck that they form a hood; and it has,
proportionately to its size, elongated wing and tail feathers. The
trumpeter and laugher, as their names express, utter a very different
coo from the other breeds. The fantail has thirty or even forty tail-
feathers, instead of twelve or fourteen—the normal number in all
the members of the great pigeon family: these feathers are kept
expanded, and are carried so erect, that in good birds the head and
tail touch: the oil-gland is quite aborted. Several other less distinct
breeds might be specified.

In the skeletons of the several breeds, the development of the
bones of the face in length and breadth and curvature differs enormously. The shape, as well as the breadth and length of the ramus
of the lower jaw, varies in a highly remarkable manner. The caudal
and sacral vertebræ vary in number; as does the number of the
ribs, together with their relative breadth and the presence of proc-
esses. The size and shape of the apertures in the sternum are highly
variable; so is the degree of divergence and relative size of the two
arms of the furcula. The proportional width of the gape of mouth,
the proportional length of the eyelids, of the orifice of the nostrils,
of the tongue (not always in strict correlation with the length of
beak), the size of the crop and of the upper part of the œsophagus;
the development and abortion of the oil-gland; the number of the
primary wing and caudal feathers; the relative length of the wing
and tail to each other and to the body; the relative length of the leg
and foot; the number of scutellæ on the toes, the development of
skin between the toes, are all points of structure which are variable.
The period at which the perfect plumage is acquired varies, as does
the state of the down with which the nestling birds are clothed
when hatched. The shape and size of the eggs vary. The manner of
flight, and in some breeds the voice and disposition, differ remark-
ably. Lastly, in certain breeds, the males and females have come to
differ in a slight degree from each other.

Altogether at least a score of pigeons might be chosen, which, if
shown to an ornithologist, and he were told that they were wild
birds, would certainly be ranked by him as well-defined species.
Moreover, I do not believe that any ornithologist would in this case
place the English carrier, the short-faced tumbler, the runt, the
barb, pouter, and fantail in the same genus; more especially as in
each of these breeds several truly-inherited sub-breeds, or species, as
he would call them, could be shown him.

Great as are the differences between the breeds of the pigeon, I am fully convinced that the common opinion of naturalists is correct, namely, that all are descended from the rock-pigeon (Columba livia), including under this term several geographic races or sub-species, which differ from each other in the most trifling respects. * * *

From these several reasons, namely,—the improbability of man having formerly made seven or eight supposed species of pigeons to breed freely under domestication;—these supposed species being quite unknown in a wild state, and their not having become anywhere feral;—these species presenting certain very abnormal characters, as compared with all other Columbidæ, though so like the rock-pigeon in most respects;—the occasional re-appearance of the blue colour and various black marks in all the breeds, both when kept pure and when crossed;—and lastly, the mongrel offspring being perfectly fertile;—from these several reasons taken together, we may safely conclude that all our domestic breeds are descended from the rock-pigeon or Columba livia with its geographical sub-species.

In favour of this view, I may add, firstly, that the wild C. livia has been found capable of domestication in Europe and in India; and that it agrees in habits and in a great number of points of structure 'with all the domestic breeds. Secondly, that, although an English carrier or a short-faced tumbler differs immensely in certain characters from the rock-pigeon, yet that, by comparing the several sub-breeds of these two races, more especially those brought from distant countries, we can make, between them and the rock-pigeon, an almost perfect series; so we can in some other cases, but not with all the breeds. Thirdly, those characters which are mainly distinctive of each breed are in each eminently variable, for instance the wattle and length of beak of the carrier, the shortness of that of the tumbler, and the number of tail-feathers in the fantail; and the explanation of this fact will be obvious when we treat of Selection. Fourthly, pigeons have been watched and tended with the utmost care, and loved by many people. They have been domesticated for thousands of years in several quarters of the world * * * The paramount importance of these considerations in explaining the immense amount of variation which pigeons have undergone, will likewise be obvious when we treat of Selection. We shall then, also, see how it is that the several breeds so often have a somewhat monstrous character. It is also a most favourable circumstance for the production of distinct breeds, that male and female pigeons can be easily mated for life; and thus different breeds can be kept together in the same aviary.

I have discussed the probable origin of domestic pigeons at

some, yet quite insufficient, length; because when I first kept pigeons and watched the several kinds, well knowing how truly they breed, I felt fully as much difficulty in believing that since they had been domesticated they had all proceeded from a common parent, as any naturalist could in coming to a similar conclusion in regard to the many species of finches, or other groups of birds, in nature. One circumstance has struck me much; namely, that nearly all the breeders of the various domestic animals and the cultivators of plants, with whom I have conversed, or whose treatises I have read, are firmly convinced that the several breeds to which each has attended, are descended from so many aboriginally distinct species. Ask, as I have asked, a celebrated raiser of Hereford cattle, whether his cattle might not have descended from Long-horns, or both from a common parent-stock, and he will laugh you to scorn. I have never met a pigeon, or poultry, or duck, or rabbit fancier, who was not fully convinced that each main breed was descended from a distinct species. Van Mons, in his treatise on pears and apples, shows how utterly he disbelieves that the several sorts, for instance a Ribston-pippin or Codlin-apple, could ever have proceeded from the seeds of the same tree. Innumerable other examples could be given. The explanation, I think, is simple: from long-continued study they are strongly impressed with the differences between the several races; and though they well know that each race varies slightly, for they win their prizes by selecting such slight differences, yet they ignore all general arguments, and refuse to sum up in their minds slight differences accumulated during many successive generations. May not those naturalists who, knowing far less of the laws of inheritance than does the breeder, and knowing no more than he does of the intermediate links in the long lines of descent, yet admit that many of our domestic races are descended from the same parents—may they not learn a lesson of caution, when they deride the idea of species in a state of nature being lineal descendants of other species?

PRINCIPLES OF SELECTION ANCIENTLY FOLLOWED, AND THEIR EFFECTS

Let us now briefly consider the steps by which domestic races have been produced, either from one or from several allied species. Some effect may be attributed to the direct and definite action of the external conditions of life, and some to habit; but he would be a bold man who would account by such agencies for the differences between a dray- and race-horse, a greyhound and bloodhound, a carrier and tumbler pigeon. One of the most remarkable features in our domesticated races is that we see in them adaptation, not indeed to the animal's or plant's own good, but to man's use or fancy. Some variations useful to him have probably arisen suddenly, or by one step; many botanists, for instance, believe that the fuller's teasel, with its hooks, which cannot be rivalled by any mechanical contrivance, is only a variety of the wild Dipsacus; and this amount of change may have suddenly arisen in a seedling. So it has probably been with the turnspit dog; and this is known to have been the case with the ancon sheep. But when we compare the dray-horse and race-horse, the dromedary and camel, the various breeds of sheep fitted either for cultivated land or mountain pasture, with the wool of one breed good for one purpose, and that of another breed for another purpose; when we compare the many breeds of dogs, each good for man in different ways; when we compare the game-cock, so pertinacious in battle, with other breeds so little quarrelsome, with "everlasting layers" which never desire to sit, and with the bantam so small and elegant; when we compare the host of agricultural, culinary, orchard, and flower-garden races of plants, most useful to man at different seasons and for different purposes, or so beautiful in his eyes, we must, I think, look further than to mere variability. We cannot suppose that all the breeds were suddenly produced as perfect and as useful as we now see them; indeed, in many cases, we know that this has not been their history. The key is man's power of accumulative selection: nature gives successive variations; man adds them up in certain directions useful to him. In this sense he may be said to have made for himself useful breeds. * * *

At the present time, eminent breeders try by methodical selection, with a distinct object in view, to make a new strain or sub-breed, superior to anything of the kind in the country. But, for our purpose, a form of Selection, which may be called Unconscious, and which results from every one trying to possess and breed from the best individual animals, is more important. Thus, a man who intends keeping pointers naturally tries to get as good dogs as he can, and afterwards breeds from his own best dogs, but he has no wish or expectation of permanently altering the breed. Nevertheless we may infer that this process, continued during centuries, would improve and modify any breed, in the same way as Bakewell, Collins, &c., by this very same process, only carried on more methodically, did greatly modify, even during their lifetimes, the forms and qualities of their cattle. Slow and insensible changes of this kind can never be recognised unless actual measurements or careful drawings of the breeds in question have been made long ago, which may serve for comparison. In some cases, however, unchanged, or but little changed individuals of the same breed exist in less civilised districts, where the breed has been less improved. There is reason to believe that King Charles's spaniel has been unconsciously modified to a large extent since the time of that monarch. Some highly competent authorities are convinced that the setter is directly derived from the spaniel, and has probably been slowly altered from it. It is known that the English pointer has been greatly changed within the last century, and in this case the change has, it is believed, been chiefly effected by crosses with the foxhound; but what concerns us is, that the change has been effected unconsciously and gradually, and yet so effectually, that, though the old Spanish pointer certainly came from Spain, Mr. Borrow has not seen, as I am informed by him, any native dog in Spain like our pointer.

By a similar process of selection, and by careful training, English racehorses have come to surpass in fleetness and size the parent Arabs, so that the latter, by the regulations for the Goodwood Races, are favoured in the weights which they carry. Lord Spencer and others have shown how the cattle of England have increased in weight and in early maturity, compared with the stock formerly kept in this country. By comparing the accounts given in various old treatises of the former and present state of carrier and tumbler pigeons in Britain, India, and Persia, we can trace the stages through which they have insensibly passed, and come to differ so greatly from the rock-pigeon. * * *

CIRCUMSTANCES FAVOURABLE TO MAN'S POWER OF SELECTION

* * * To sum up on the origin of our domestic races of animals and plants, Changed conditions of life are of the highest importance in causing variability, both by acting directly on the organisation, and indirectly by affecting the reproductive system. It is not probable that variability is in inherent and necessary contingent, under all circumstances. The greater or less force of inheritance and reversion, determine whether variations shall endure. Variability is governed by many unknown laws, of which correlated growth is probably the most important. Something, but how much we do not know, may be attributed to the definite action of the conditions of life. Some, perhaps a great, effect may be attributed to the increased use or disuse of parts. The final result is thus rendered infinitely complex. In some cases the intercrossing of aboriginally distinct species appears to have played an important part in the origin of our breeds. When several breeds have once been formed in any country, their occasional intercrossing, with the aid of selection, has, no doubt, largely aided in the formation of new sub-breeds; but the importance of crossing has been much exaggerated, both in regard to animals and to those plants which are propagated by seed. With plants which are temporarily propagated by cuttings, buds, &c., the importance of crossing is immense; for the cultivator may here disregard the extreme variability both of hybrids and of mongrels, and the sterility of hybrids; but plants not propagated by

seed are of little importance to us, for their endurance is only temporary. Over all these causes of Change, the accumulative action of Selection, whether applied methodically and quickly, or unconsciously and slowly but more efficiently seems to have been the predominant Power.

Chapter II

VARIATION UNDER NATURE

Variability — Individual differences — Doubtful species — Wideranging, much diffused, and common species vary most—Species of the larger genera in each country vary more frequently than the species of the smaller genera—Many of the species of the larger genera resemble varieties in being very closely, but unequally, related to each other, and in having restricted ranges.

Before applying the principles arrived at in the last chapter to organic beings in a state of nature, we must briefly discuss whether these latter are subject to any variation. To treat this subject properly, a long catalogue of dry facts ought to be given; but these I shall reserve for a future work. Nor shall I here discuss the various definitions which have been given of the term species. No one definition has satisfied all naturalists; yet every naturalist knows vaguely what he means when he speaks of a species. Generally the term includes the unknown element of a distant act of creation. The term "variety" is almost equally difficult to define; but here community of descent is almost universally implied, though it can rarely be proved. * * *

WIDE-RANGING, MUCH DIFFUSED, AND COMMON
SPECIES VARY MOST

* * * Alphonse de Candolle and others have shown that plants which have very wide ranges generally present varieties; and this might have been expected, as they are exposed to diverse physical conditions, and as they come into competition (which, as we shall hereafter see, is an equally or more important circumstance) with different sets of organic beings. But my tables further show that, in any limited country, the species which are the most common, that is abound most in individuals, and the species which are most widely diffused within their own country (and this is a different consideration from wide range, and to a certain extent from commonness), oftenest give rise to varieties sufficiently well marked to have been recorded in botanical works. Hence it is the most flourishing, or, as they may be called, the dominant species,—those which range widely, are the most diffused in their own country, and are the most numerous in individuals,—which oftenest produce well-

marked varieties, or, as I consider them, incipient species. And this, perhaps, might have been anticipated; for, as varieties, in order to become in any degree permanent, necessarily have to struggle with the other inhabitants of the country, the species which are already dominant will be the most likely to yield off-spring, which, though in some slight degree modified, still inherit those advantages that enabled their parents to become dominant over their compatriots. * * *

SUMMARY

Finally, varieties cannot be distinguished from species,—except, first, by the discovery of intermediate linking forms; and, secondly, by a certain indefinite amount of difference between them; for two forms, if differing very little, are generally ranked as varieties, notwithstanding that they cannot be closely connected; but the amount of difference considered necessary to give to any two forms the rank of species cannot be defined. In genera having more than the average number of species in any country, the species of these genera have more than the average number of varieties. In large genera the species are apt to be closely, but unequally, allied together, forming little clusters round other species. Species very closely allied to other species apparently have restricted ranges. In all these respects the species of large genera present a strong analogy with varieties. And we can clearly understand these analogies, if species once existed as varieties, and thus originated; whereas, these analogies are utterly inexplicable if species are independent creations.

We have, also, seen that it is the most flourishing or dominant species of the larger genera within each class which on an average yield the greatest number of varieties; and varieties, as we shall hereafter see, tend to become converted into new and distinct species. Thus the larger genera tend to become larger; and throughout nature the forms of life which are now dominant tend to become still more dominant by leaving many modified and dominant descendants. But by steps hereafter to be explained, the larger genera also tend to break up into smaller genera. And thus, the forms of life throughout the universe become divided into groups subordinate to groups.

Chapter III

STRUGGLE FOR EXISTENCE

Its bearing on natural selection—The term used in a wide sense—Geometrical ratio of increase—Rapid increase of naturalised ani-

*mals and plants—Nature of the checks in increase—Competition
universal—Effects of climate—Protection from the number of individuals—Complex relations of all animals and plants throughout
nature—Struggle for life most severe between individuals and varieties of the same species: often severe between species of the
same genus—The relation of organism to organism the most important of all relations.*

Before entering on the subject of this chapter, I must make a
few preliminary remarks, to show how the struggle for existence
bears on Natural Selection. It has been seen in the last chapter
that amongst organic beings in a state of nature there is some individual variability: indeed I am not aware that this has ever been
disputed. It is immaterial for us whether a multitude of doubtful
forms be called species or sub-species or varieties; what rank, for
instance, the two or three hundred doubtful forms of British plants
are entitled to hold, if the existence of any well-marked varieties
be admitted. But the mere existence of individual variability and
of some few well-marked varieties, though necessary as the foundation for the work, helps us but little in understanding how species
arise in nature. How have all those exquisite adaptations of one
part of the organisation to another part, and to the conditions of
life, and of one organic being to another being, been perfected?
We see these beautiful co-adaptations most plainly in the woodpecker and the mistletoe; and only a little less plainly in the hum-
blest parasite which clings to the hairs of a quadruped or feathers
of a bird; in the structure of the beetle which dives through the
water; in the plumed seed which is wafted by the gentlest breeze; in
short, we see beautiful adaptations everywhere and in every part of
the organic world.

Again, it may be asked, how is it that varieties, which I have
called incipient species, become ultimately converted into good
and distinct species which in most cases obviously differ from each
other far more than do the varieties of the same species? How do
those groups of species, which constitute what are called distinct
genera, and which differ from each other more than do the species
of the same genus, arise? All these results, as we shall more fully see
in the next chapter, follow from the struggle for life. Owing to this
struggle, variations, however slight and from whatever cause proceeding, if they be in any degree profitable to the individuals of a
species, in their infinitely complex relations to other organic beings
and to their physical conditions of life, will tend to the preservation
of such individuals, and will generally be inherited by the offspring. The offspring, also, will thus have a better chance of surviving, for, of the many individuals of any species which are

periodically born, but a small number can survive. I have called this principle, by which each slight variation, if useful, is preserved, by the term Natural Selection, in order to mark its relation to man's power of selection. But the expression often used by Mr. Herbert Spencer of the Survival of the Fittest is more accurate, and is sometimes equally convenient. We have seen that man by selection can certainly produce great results, and can adapt organic beings to his own uses, through the accumulation of slight but useful variations, given to him by the hand of Nature. But Natural Selection, as we shall hereafter see, is a power incessantly ready for action, and is as immeasurably superior to man's feeble efforts, as the works of Nature are to those of Art.

We will now discuss in a little more detail the struggle for existence. In my future work this subject will be treated, as it well deserves, at greater length. The elder De Candolle and Lyell have largely and philosophically shown that all organic beings are exposed to severe competition. In regard to plants, no one has treated this subject with more spirit and ability than W. Herbert, Dean of Manchester, evidently the result of his great horticultural knowledge. Nothing is easier than to admit in words the truth of the universal struggle for life, or more difficult—at least I have found it so—than constantly to bear this conclusion in mind. Yet unless it be thoroughly engrained in the mind, the whole economy of nature, with every fact on distribution, rarity, abundance, extinction, and variation, will be dimly seen or quite misunderstood. We behold the face of nature bright with gladness, we often see superabundance of food; we do not see or we forget, that the birds which are idly singing round us mostly live on insects or seeds, and are thus constantly destroying life; or we forget how largely these songsters, or their eggs, or their nestlings, are destroyed by birds and beasts of prey; we do not always bear in mind, that, though food may be now superabundant, it is not so at all seasons of each recurring year.

THE TERM, STRUGGLE FOR EXISTENCE, USED IN A LARGE SENSE

I should premise that I use this term in a large and metaphorical sense including dependence of one being on another, and including (which is more important) not only the life of the individual, but success in leaving progeny. Two canine animals, in a time of dearth, may be truly said to struggle with each other which shall get food and live. But a plant on the edge of a desert is said to struggle for life against the drought, though more properly it should be said to be dependent on the moisture. A plant which annually produces a thousand seeds, of which only one of an average comes to maturity, may be more truly said to struggle with the plants of the same and other kinds which already clothe the ground. The mistletoe is dependent on the apple and a few other trees, but can only in a far-fetched sense be said to struggle with these trees, for, if too many of these parasites grow on the same tree, it languishes and dies. But several seedling mistletoes, growing close together on the same branch, may more truly be said to struggle with each other. As the mistletoe is disseminated by birds, its existence depends on them; and it may methodically be said to struggle with other fruit-bearing plants, in tempting the birds to devour and thus disseminate its seeds. In these several senses, which pass into each other, I use for convenience' sake the general term of Struggle for Existence.

GEOMETRICAL RATIO OF INCREASE

A struggle for existence inevitably follows from the high rate at which all organic beings tend to increase. Every being, which during its natural lifetime produces several eggs or seeds, must suffer destruction during some period of its life, and during some season or occasional year, otherwise, on the principle of geometrical increase, its numbers would quickly become so inordinately great that no country could support the product. Hence, as more individuals are produced than can possibly survive, there must in every case be a struggle for existence, either one individual with another of the same species, or with the individuals of distinct species, or with the physical conditions of life. It is the doctrine of Malthus applied with manifold force to the whole animal and vegetable kingdoms; for in this case there can be no artificial increase of food, and no prudential restraint from marriage. Although some species may be now increasing, more or less rapidly, in numbers, all cannot do so, for the world would not hold them.

There is no exception to the rule that every organic being naturally increases at so high a rate, that, if not destroyed, the earth would soon be covered by the progeny of a single pair. Even slow-breeding man has doubled in twenty-five years, and at this rate, in less than a thousand years, there would literally not be standing-room for his progeny. Linnæus has calculated that if an annual plant produced only two seeds—and there is no plant so unproductive as this—and their seedlings next year produced two, and so on, then in twenty years there should be a million plants. The elephant is reckoned the slowest breeder of all known animals, and I have taken some pains to estimate its probable minimum rate of natural increase; it will be safest to assume that it begins breeding when thirty years old, and goes on breeding till ninety years old, bringing forth six young in the interval, and surviving till one hundred years old; if this be so, after a period of from 740 to 750 years there would be nearly nineteen million elephants alive, descended from the first pair.

But we have better evidence on this subject than mere theoretical calculations, namely, the numerous recorded cases of the astonishingly rapid increase of various animals in a state of nature, when circumstances have been favourable to them during two or three following seasons. Still more striking is the evidence from our domestic animals of many kinds which have run wild in several parts of the world; if the statements of the rate of increase of slow-breeding cattle and horses in South America, and latterly in Australia, had not been well authenticated, they would have been incredible. So it is with plants; cases could be given of introduced plants which have become common throughout whole islands in a period of less than ten years. Several of the plants, such as the cardoon and a tall thistle, which are now the commonest over the whole plains of La Plata, clothing square leagues of surface almost to the exclusion of every other plant, have been introduced from Europe; and there are plants which now range in India, as I hear from Dr. Falconer, from Cape Comorin to the Himalaya, which have been imported from America since its discovery. In such cases, and endless others could be given, no one supposes, that the fertility of the animals or plants has been suddenly and temporarily increased in any sensible degree. The obvious explanation is that the conditions of life have been highly favourable, and that there has consequently been less destruction of the old and young, and that nearly all the young have been enabled to breed. Their geometrical ratio of increase, the result of which never fails to be surprising, simply explains their extraordinarily rapid increase and wide diffusion in their new homes.

In a state of nature almost every full-grown plant annually produces seed, and amongst animals there are very few which do not annually pair. Hence we may confidently assert, that all plants and animals are tending to increase at a geometrical ratio,—that all would rapidly stock every station in which they could anyhow exist, —and that this geometrical tendency to increase must be checked by destruction at some period of life. Our familiarity with the larger domestic animals tends, I think, to mislead us: we see no great destruction falling on them, but we do not keep in mind that thousands are annually slaughtered for food, and that in a state of nature an equal number would have somehow to be disposed of.

The only difference between organisms which annually produce eggs or seeds by the thousand, and those which produce extremely few, is, that the slow-breeders would require a few more years to people, under favourable conditions, a whole district, let it be ever so large. The condor lays a couple of eggs and the ostrich a score, and yet in the same country the condor may be the more numerous of the two; the Fulmar petrel lays but one egg, yet it is believed to be the most numerous bird in the world. One fly deposits hundreds of eggs, and another, like the hippobosca, a single one; but this difference does not determine how many individuals of the two species can be supported in a district. A large number of eggs is of some importance to those species which depend on a fluctuating amount of food, for it allows them rapidly to increase in number. But the real importance of a large number of eggs or seeds is to make up for much destruction at some period of life; and this period in the great majority of cases is an early one. If an animal can in any way protect its own eggs or young, a small number may be produced, and yet the average stock be fully kept up; but if many eggs or young are destroyed, many must be produced, or the species will become extinct. It would suffice to keep up the full number of a tree, which lived on an average for a thousand years, if a single seed were produced once in a thousand years, supposing that this seed were never destroyed, and could be ensured to germinate in a fitting place. So that, in all cases, the average number of any animal or plant depends only indirectly on the number of its eggs or seeds.

In looking at Nature, it is most necessary to keep the foregoing considerations always in mind—never to forget that every single organic being may be said to be striving to the utmost to increase in numbers; that each lives by a struggle at some period of its life; that heavy destruction inevitably falls either on the young or old, during each generation or at recurrent intervals. Lighten any check, mitigate the destruction ever so little, and the number of the species will almost instantaneously increase to any amount. * * *

Chapter IV

NATURAL SELECTION; OR THE SURVIVAL OF THE FITTEST

Natural Selection—its power compared with man's selection—its power on characters of trifling importance—its power at all ages and on both sexes—Sexual selection—On the generality of intercrosses between individuals of the same species—Circumstances favourable and unfavourable to the results of natural selection, namely, intercrossing, isolation, number of individuals—Slow action—Extinction caused by natural selection—Divergence of character, related to the diversity of inhabitants of any small area, and to naturalisation—Action of natural selection, through divergence of character and extinction, on the descendants from a common parent—Explains the grouping of all organic beings—Advance in organisation—Low forms preserved—Convergence of character—Indefinite multiplication of species—Summary.

How will the struggle for existence, briefly discussed in the last chapter, act in regard to variation? Can the principle of selection, which we have seen is so potent in the hands of man, apply under nature? I think we shall see that it can act most efficiently. Let the endless number of slight variations and individual differences occurring in our domestic productions, and, in a lesser degree, in those under nature, be borne in mind; as well as the strength of the hereditary tendency. Under domestication, it may be truly said that the whole organisation becomes in some degree plastic. But the variability, which we almost universally meet with in our domestic productions, is not directly produced, as Hooker and Asa Gray have well remarked, by man; he can neither originate varieties, nor prevent their occurrence; he can preserve and accumulate such as do occur. Unintentionally he exposes organic beings to new and changing conditions of life, and variability ensues; but similar changes of conditions might and do occur under nature. Let it also be borne in mind how infinitely complex and close-fitting are the mutual relations of all organic beings to each other and to their physical conditions of life; and consequently what infinitely varied diversities of structure might be of use to each being under changing conditions of life. Can it, then, be thought improbable, seeing that variations useful to man have undoubtedly occurred, that other variations useful in some way to each being in the great and complex battle of life, should occur in the course of many successive generations. If such do occur, can we doubt (remembering that many more individuals are born than can possibly survive) that individuals having any advantage, however slight, over others, would have the best chance of surviving and of procreating their kind? On the other hand, we may feel sure that any variation in the least degree injurious would be rigidly destroyed. This preservation of favourable individual differences and variations, and the destruction of those which are injurious, I have called Natural Selection, or the Survival of the Fittest. Variations neither useful nor injurious would not be affected by natural selection, and would be left either a fluctuating element, as perhaps we see in certain polymorphic species, or would ultimately become fixed, owing to the nature of the organism and the nature of the conditions.

Several writers have misapprehended or objected to the term Natural Selection. Some have even imagined that natural selection induces variability, whereas it implies only the preservation of such variations as arise and are beneficial to the being under its conditions of life. No one objects to agriculturists speaking of the potent effects of man's selection; and in this case the individual differences given by nature, which man for some object selects, must of necessity first occur. Others have objected that the term selection implies conscious choice in the animals which become modified; and it had even been urged that, as plants have no volition, natural selection is not applicable to them! In the literal sense of the word, no doubt, natural selection is a false term; but who ever objected to chemists speaking of the elective affinities of the various elements?—and yet an acid cannot strictly be said to elect the base with which it in preference combines. It has been said that I speak of natural selection as an active power or Deity; but who objects to an author speaking of the attraction of gravity as ruling the movements of the planets? Every one knows what is meant and is implied by such metaphorical expressions; and they are almost necessary for brevity. So again it is difficult to avoid personifying the word Nature; but I mean by Nature, only the aggregate action and product of many natural laws, and by laws the sequence of events as ascertained by us. With a little familiarity such superficial objections will be forgotten.

We shall best understand the probable course of natural selection by taking the case of a country undergoing some slight physical change, for instance, of climate. The proportional numbers of its inhabitants will almost immediately undergo a change, and some species will probably become extinct. We may conclude, from what we have seen of the intimate and complex manner in which the inhabitants of each country are bound together, that any change in the numerical proportions of the inhabitants, independently of the change of climate itself, would seriously affect the others. If the country were open on its borders, new forms would certainly immigrate, and this would likewise seriously disturb the relations of some of the former inhabitants. Let it be remembered how powerful the influence of a single introduced tree or mammal has been shown to be. But in the case of an island, or of a country partly surrounded by barriers, into which new and better adapted forms could not freely enter, we should then have places in the economy of nature which would assuredly be better filled up, if some of the original inhabitants were in some manner modified; for, had the area been open to immigration, these same places would have been seized on by intruders. In such cases, slight modifications, which in any way favoured the individuals of any species, by better adapting them to their altered conditions, would tend to be preserved; and natural selection would have free scope for the work of improvement.

We have good reason to believe, as shown in the first chapter, that changes in the conditions of life give a tendency to increased variability; and in the foregoing cases the conditions have changed, and this would manifestly be favourable to natural selection, by affording a better chance of the occurrence of profitable variations. Unless such occur, natural selection can do nothing. Under the term of "variations," it must never be forgotten that mere individual differences are included. As man can produce a great result with his domestic animals and plants by adding up in any given direction individual differences, so could natural selection, but far more easily from having incomparably longer time for action. Nor do I believe that any great physical change, as of climate, or any unusual degree of isolation to check immigration, is necessary in order that new and unoccupied places should be left, for natural selection to fill up by improving some of the varying inhabitants. For as all the inhabitants of each country are struggling together with nicely balanced forces, extremely slight modifications in the structure or habits of one species would often give it an advantage over others; and still further modifications of the same kind would often still further increase the advantage, as long as the species continued under the same conditions of life and profited by similar means of subsistence and defence. No country can be named in which all the native inhabitants are now so perfectly adapted to each other and to the physical conditions under which they live, that none of them could be still better adapted or improved; for in all countries, the natives have been so far conquered by naturalised productions, that they have allowed some foreigners to take firm possession of the land. And as foreigners have thus in every country beaten some of the natives, we may safely conclude that the natives might have been modified with advantage, so as to have better resisted the intruders.

As man can produce, and certainly has produced, a great result by his methodical and unconscious means of selection, what may not natural selection effect? Man can act only on external and visible characters: Nature, if I may be allowed to personify the natural preservation or survival of the fittest, cares nothing for appearances, except in so far as they are useful to any being. She can act on every internal organ, on every shade of constitutional difference, on the whole machinery of life. Man selects only for his own good: Nature only for that of the being which she tends. Every selected character is fully exercised by her, as is implied by the fact of their selection. Man keeps the natives of many climates in the same country; he seldom exercises each selected character in some peculiar and fitting manner; he feeds a long and a short beaked pigeon on the same food; he does not exercise a long-backed or long-legged quadruped in any peculiar manner; he exposes sheep with long and short wool to the same climate. He does not allow the most vigorous males to struggle for the females. He does not rigidly destroy all inferior animals, but protects during each varying season, as far as lies in his power, all his productions. He often begins his selection by some half-monstrous form; or at least by some modification prominent enough to catch the eye or to be plainly useful to him. Under Nature, the slightest differences of structure or constitution may well turn the nicely balanced scale in the struggle for life, and so be preserved. How fleeting are the wishes and efforts of man! how short his time! and consequently how poor will be his results, compared with those accumulated by Nature during whole geological periods! Can we wonder, then, that Nature's productions should be far "truer" in character than man's productions that they should be infinitely better adapted to the most complex conditions of life and should plainly bear the stamp of far higher workmanship?

It may metaphorically be said that natural selection is daily and hourly scrutinising, throughout the world, the slightest variations; rejecting those that are bad, preserving and adding up all that are good; silently and insensibly working, *whenever and wherever opportunity offers*, at the improvement of each organic being in relation to its organic and inorganic conditions of life. We see nothing of these slow changes in progress, until the hand of time has marked the lapse of ages, and then so imperfect is our view into long-past geological ages, that we see only that the forms of life are now different from what they formerly were.

In order that any great amount of modification should be effected in a species, a variety when once formed must again, perhaps after a long interval of time, vary or present individual differences of the same favourable nature as before; and these must be again preserved, and so onwards step by step. Seeing that individual differences of the same kind perpetually recur, this can hardly be considered as an unwarrantable assumption. But whether it is true, we can judge only by seeing how far the hypothesis accords with and explains the general phenomena of nature. On the other hand, the ordinary belief that the amount of possible variation is a strictly limited quantity is likewise a simple assumption.

Although natural selection can act only through and for the good of each being, yet characters and structures, which we are apt to consider as of very trifling importance, may thus be acted on. When we see leaf-eating insects green, and bark-feeders mottled-grey; the alpine ptarmigan white in winter, the red-grouse the colour of heather, we must believe that these tints are of service to these birds and insects in preserving them from danger. Grouse, if not destroyed at some period of their lives would increase in countless numbers; they are known to suffer largely from birds of prey; and hawks are guided by eyesight to their prey—so much so, that on parts of the Continent persons are warned not to keep white pigeons, as being the most liable to destruction. Hence natural selection might be effective in giving the proper colour to each kind of grouse, and in keeping that colour, when once acquired, true and constant. Nor ought we to think that the occasional destruction of an animal of any particular colour would produce little effect: we should remember how essential it is in a flock of white sheep to destroy a lamb with the faintest trace of black. We have seen how the colour of the hogs, which feed on the "paint-root" in Virginia, determines whether they shall live or die. In plants, the down on the fruit and the colour of the flesh are considered by botanists as characters of the most trifling importance: yet we hear from an excellent horticulturist, Downing, that in the United States, smooth-skinned fruits suffer far more from a beetle, a Curculio, than those with down; that purple plums suffer far more from a certain disease than yellow plums; whereas another disease attacks yellow-fleshed peaches far more than those with other coloured flesh. If, with all the aids of art, these slight differences make a great difference in cultivating the several varieties, assuredly, in a state of nature, where the trees would have to struggle with other trees, and with a host of enemies, such differences would effectually settle which variety, whether a smooth or downy, a yellow or purple fleshed fruit, should succeed.

In looking at many small points of difference between species, which, as far as our ignorance permits us to judge, seem quite unimportant, we must not forget that climate, food, &c., have no doubt produced some direct effect. It is also necessary to bear in mind that, owing to the law of correlation, when one part varies, and the variations are accumulated through natural selection, other modifications, often of the most unexpected nature, will ensue.

As we see that those variations which, under domestication, appear at any particular period of life, tend to reappear in the offspring at the same period;—for instance, in the shape, size, and flavour of the seeds of the many varieties of our culinary and agricultural plants; in the caterpillar and cocoon stages of the varieties of the silk-worm; in the eggs of poultry, and in the colour of the down of their chickens; in the horns of our sheep and cattle when nearly adult;—so in a state of nature natural selection will be enabled to act on and modify organic beings at any age, by the accumulation of variations profitable at that age, and by their inheritance at a corresponding age. If it profit a plant to have its seeds more and more widely disseminated by the wind, I can see no greater difficulty in this being effected through natural selection, than in the cotton-planter increasing and improving by selection the down in the pods on his cotton-trees. Natural selection may modify and adapt the larva of an insect to a score of contingencies, wholly different from those which concern the mature insect; and these modifications may effect, through correlation, the structure of the adult. So, conversely, modifications in the adult may affect the structure of the larva; but in all cases natural selection will ensure that they shall not be injurious: for if they were so, the species would become extinct.

Natural selection will modify the structure of the young in relation to the parent, and of the parent in relation to the young. In social animals it will adapt the structure of each individual for the benefit of the whole community; if the community profits by the selected change. What natural selection cannot do, is to modify the structure of one species, without giving it any advantage, for the good of another species; and though statements to this effect may be found in works of natural history, I cannot find one case which will bear investigation. A structure used only once in an animal's life, if of high importance to it, might be modified to any extent by natural selection; for instance, the great jaws possessed by certain insects, used exclusively for opening the cocoon—or the hard tip to the beak of unhatched birds, used for breaking the egg. It has been asserted, that of the best short-beaked tumbler-pigeons a greater number perish in the egg than are able to get out of it; so that fanciers assist in the act of hatching. Now if nature had to make the beak of a full-grown pigeon very short for the bird's own advantage, the process of modification would be very slow, and there would be simultaneously the most rigorous selection of all the young birds within the egg, which had the most powerful and hardest beaks, for all with weak beaks would inevitably perish; or, more delicate and more easily broken shells might be selected, the thickness of the shell being known to vary like every other structure.

It may be well here to remark that with all beings there must be much fortuitous destruction, which can have little or no influence on the course of natural selection. For instance a vast number of eggs or seeds are annually devoured, and these could be modified through natural selection only if they varied in some manner which protected them from their enemies. Yet many of these eggs or seeds would perhaps, if not destroyed, have yielded individuals better adapted to their conditions of life than any of those which happened to survive. So again a vast number of mature animals and plants, whether or not they be the best adapted to their conditions, must be annually destroyed by accidental causes, which would not be in the least degree mitigated by certain changes of structure or constitution which would in other ways be beneficial to the species. But let the destruction of the adults be ever so heavy, if the number which can exist in any district be not wholly kept down by such causes,—or again let the destruction of eggs or seeds be so great that only a hundredth or a thousandth part are developed,—yet of those which do survive, the best adapted individuals, supposing that there is any variability in a favourable direction, will tend to propagate their kind in larger numbers than the less well adapted. If the numbers be wholly kept down by the causes just indicated, as will often have been the case, natural selection will be powerless in certain beneficial directions; but this is no valid objection to its efficiency at other times and in other ways; for we are far from having any reason to suppose that many species ever undergo modification and improvement at the same time in the same area.

SEXUAL SELECTION

Inasmuch as peculiarities often appear under domestication in one sex and become hereditarily attached to that sex, so no doubt it will be under nature. Thus it is rendered possible for the two sexes to be modified through natural selection in relation to different habits of life, as is sometimes the case; or for one sex to be modified in relation to the other sex, as commonly occurs. This leads me to say a few words on what I have called Sexual Selection. This form of selection depends, not on a struggle for existence in relation to other organic beings or to external conditions, but on a struggle between the individuals of one sex, generally the males, for the possession of the other sex. The result is not death to the unsuccessful competitor, but few or no offspring. Sexual selection is, therefore, less rigorous than natural selection. Generally, the most vigorous males, those which are best fitted for their places in nature, will leave most progeny. But in many cases, victory depends not so much on general vigor, as on having special weapons, confined to the male sex. A hornless stag or spurless cock would have poor chance of leaving numerous offspring. Sexual selection, by always allowing the victor to breed, might surely give indomitable courage, length to the spur, and strength to the wing to strike in the spurred leg, in nearly the same manner as does the brutal cock-fighter by the careful selection of his best cocks. How low in the scale of nature the law of battle descends, I know not; male alligators have been described as fighting, bellowing, and whirling round, like Indians in a war-dance, for the possession of the females; male salmons have been observed fighting all day long; male stag-beetles sometimes bear wounds from the huge mandibles of other males; the males of certain hymenopterous insects have been frequently seen by that inimitable observer M. Fabre, fighting for a particular female who sits by, an apparently unconcerned beholder of the struggle, and then retires with the conqueror. The war is, perhaps, severest between the males of polygamous animals, and these seem oftenest provided with special weapons. The males of carnivorous animals are already well armed; though to them and to others, special means of defence may be given through means of sexual selection, as the mane of the lion, and the hooked jaw to the male salmon; for the shield may be as important for victory, as the sword or spear.

Amongst birds, the contest is often of a more peaceful character. All those who have attended to the subject, believe that there is the severest rivalry between the males of many species to attract, by singing, the females. The rock-thrush of Guiana, birds of paradise, and some others, congregate; and successive males display with the most elaborate care, and show off in the best manner, their gorgeous plumage; they likewise perform strange antics before the females, which, standing by as spectators, at last choose the most attractive partner. Those who have closely attended to birds in confinement well know that they often take individual preferences and dislikes: thus Sir R. Heron has described how a pied peacock was eminently attractive to all his hen birds. I cannot here enter on the necessary details; but if man can in a short time give beauty and an elegant carriage to his bantams, according to the standard of beauty, I can see no good reason to doubt that female birds, by selecting, during thousands of generations, the most melodious or beautiful males according to their standard of beauty, might produce a marked effect. Some well-known laws, with respect to the plumage of male and female birds, in comparison with the plumage of the young, can partly be explained through the action of sexual selection on variations occurring at different ages, and transmitted to the males alone or to both sexes at corresponding ages; but I have not space here to enter on this subject.

Thus it is, as I believe, that when the males and females of any animal have the same general habits of life, but differ in structure, colour, or ornament, such differences have been mainly caused by sexual selection: that is, by individual males having had, in successive generations, some slight advantage over other males, in their weapons, means of defence, or charms, which they have transmitted to their male offspring alone. Yet, I would not wish to attribute all sexual differences to this agency: for we see in our domestic animals peculiarities arising and becoming attached to the male sex, which apparently have not been augmented through selection by man. The tuft of hair on the breast of the wild turkey-cock cannot be of any use, and it is doubtful whether it can be ornamental in the eyes of the female bird;—indeed, had the tuft appeared under domestication, it would have been called a monstrosity.

ILLUSTRATIONS OF THE ACTION OF NATURAL SELECTION, OR THE SURVIVAL OF THE FITTEST

In order to make it clear how, as I believe, natural selection acts, I must beg permission to give one or two imaginary illustrations. Let us take the case of a wolf, which preys on various animals, securing some by craft, some by strength and some by fleetness; and let us suppose that the fleetest prey, a deer for instance, had from any change in the country increased in numbers or that other prey had decreased in numbers, during that season of the year when the wolf was hardest pressed for food. Under such circumstances the swiftest and slimmest wolves would have the best chance of surviving and so be preserved or selected,—provided always that they retained strength to master their prey at this or some other period of the year, when they were compelled to prey on other animals. I can see no more reason to doubt that this would be the result, than that man should be able to improve the fleetness of his greyhounds by careful and methodical selection, or by that kind of unconscious selection which follows from each man trying to keep the best dogs without any thought of modifying the breed. I may add, that, according to Mr. Pierce, there are two varieties of the wolf inhabiting the Catskill Mountains, in the United States, one with a light greyhound-like form, which pursues deer, and the other more bulky, with shorter legs, which more frequently attacks the shepherd's flocks.

It should be observed that, in the above illustration, I speak of the slimmest individual wolves, and not of any single strongly-marked variation having been preserved. In former editions of this work I sometimes spoke as if this latter alternative had frequently occurred. I saw the great importance of individual differences, and this led me fully to discuss the results of unconscious selection by man, which depends on the preservation of all the more or less valuable individuals, and on the destruction of the worst. I saw, also, that the preservation in a state of nature of any occasional deviation of structure, such as a monstrosity, would be a rare event; and that, if at first preserved, it would generally be lost by subsequent intercrossing with ordinary individuals. Nevertheless, until reading an able and valuable article in the 'North British Review' (1867), I did not appreciate how rarely single variations, whether slight or strongly marked, could be perpetuated. The author takes the case of a pair of animals, producing during their lifetime two hundred offspring, of which, from various causes of destruction, only two on an average survive to procreate their kind. This is rather an extreme estimate for most of the higher animals, but by no means so for many of the lower organisms. He then shows that if a single individual were born, which varied in some manner, giving it twice as good a chance of life as that of the other individuals, yet the chances would be strongly against its survival. Supposing it to survive and to breed, and that half its young inherited the favourable variation; still, as the Reviewer goes on to show, the young would have only a slightly better chance of surviving and breeding; and this chance would go on decreasing in the succeeding generations. The justice of these remarks cannot, I think, be disputed. If, for instance, a bird of some kind could procure its food more easily by having its beak curved, and if one were born with its beak strongly curved, and which consequently flourished, nevertheless there would be a very poor chance of this one individual perpetuating its kind to the exclusion of the common form; but there can hardly be a doubt, judging by what we see taking place under domestication, that this result would follow from the preservation during many generations of a large number of individuals with more or less strongly curved beaks, and from the destruction of a still larger number with the straightest beaks.

It should not, however, be overlooked that certain rather strongly marked variations, which no one would rank as mere individual differences, frequently recur owing to a similar organisation being similarly acted on—of which fact numerous instances could be given with our domestic productions. In such cases, if the varying individual did not actually transmit to its offspring its newly-acquired character, it would undoubtedly transmit to them, as long as the existing conditions remained the same, a still stronger tendency to vary in the same manner. There can also be little doubt that the tendency to vary in the same manner has often been so strong that all the individuals of the same species have been similarly modified without the aid of any form of selection. Or only a third, fifth, or tenth part of the individuals may have been thus affected, of which fact several instances could be given. Thus Graba estimates that about one-fifth of the guillemots in the Faroe Islands consist of a variety so well marked, that it was formerly ranked as a distinct species under the name of Uria lacrymans. In cases of this kind, if the variation were of a beneficial nature, the original form would soon be supplanted by the modified form, through the survival of the fittest.

To the effects of intercrossing in eliminating variations of all kinds, I shall have to recur; but it may be here remarked that most animals and plants keep to their proper homes, and do not needlessly wander about; we see this even with migratory birds, which almost always return to the same spot. Consequently each newly-formed variety would generally be at first local, as seems to be the common rule with varieties in a state of nature; so that similarly modified individuals would soon exist in a small body together, and would often breed together. If the new variety were successful in its battle for life, it would slowly spread from a central district, competing with and conquering the unchanged individuals on the margins of an ever-increasing circle.

It may be worth while to give another and more complex illustration of the action of natural selection. Certain plants excrete sweet juice, apparently for the sake of eliminating something injurious from the sap: this is effected, for instance, by glands at the base of the stipules in some Leguminosæ, and at the backs of the leaves of the common laurel. This juice, though small in quantity, is greedily sought by insects; but their visits do not in any way benefit the plant. Now, let us suppose that the juice or nectar was excreted from the inside of the flowers of a certain number of plants of any species. Insects in seeking the nectar would get dusted with pollen, and would often transport it from one flower to another. The flowers of two distinct individuals of the same species would thus get crossed; and the act of crossing, as can be fully proved, gives rise to vigorous seedlings which consequently would have the best chance of flourishing and surviving. The plants which produced flowers with the largest glands or nectaries, excreting most nectar, would oftenest be visited by insects, and would oftenest be crossed; and so in the long-run would gain the upper hand and form a local variety. The flowers, also, which had their stamens and pistils placed, in relation to the size and habits of the particular insects which visited them, so as to favour in any degree the transportal of the pollen, would likewise be favoured. We might have taken the case of insects visiting flowers for the sake of collecting pollen instead of nectar; and as pollen is formed for the sole purpose of fertilisation, its destruction appears to be a simple loss to the plant; yet if a little pollen were carried, at first occasionally and then habitually, by the pollen-devouring insects from flower to flower, and a cross thus effected, although nine-tenths of the pollen were destroyed it might still be a great gain to the plant to be thus robbed; and the individuals which produced more and more pollen, and had larger anthers, would be selected.

When our plant, by the above process long continued, had been rendered highly attractive to insects, they would, unintentionally on their part, regularly carry pollen from flower to flower; and that they do this effectually, I could easily show by many striking facts. I will give only one, as likewise illustrating one step in the separation of the sexes of plants. Some holly-trees bear only male flowers, which have four stamens producing a rather small quantity of pollen, and a rudimentary pistil: other holly-trees bear only female flowers; these have a full-sized pistil, and four stamens with shrivelled anthers, in which not a grain of pollen can be detected. Having found a female tree exactly sixty yards from a male tree, I put the stigmas of twenty flowers, taken from different branches, under the microscope, and on all, without exception, there were a few pollen-grains, and on some a profusion. As the wind had set for several days from the female to the male tree, the pollen could not thus have been carried. The weather had been cold and boisterous, and therefore not favourable to bees, nevertheless every female flower which I examined had been effectually fertilised by the bees, which had flown from tree to tree in search of nectar. But to return to our imaginary case: as soon as the plant had been rendered so highly attractive to insects that pollen was regularly carried from flower to flower, another process might commence. No naturalist doubts the advantage of what has been called the "physiological division of labour"; hence we may believe that it would be advantageous to a plant to produce stamens alone in one flower or on one whole plant, and pistils alone in another flower or on another plant. In plants under culture and placed under new conditions of life, sometimes the male organs and sometimes the female organs become more or less impotent; now if we suppose this to occur in ever so slight a degree under nature, then, as pollen is already carried regularly from flower to flower, and as a more complete separation of the sexes of our plant would be advantageous on the principle of the division of labour, individuals with this tendency more and more increased would be continually favoured or selected, until at last a complete separation of the sexes might be effected. It would take up too much space to show the various steps, through dimorphism and other means, by which the separation of the sexes in plants of various kinds is apparently now in progress; but I may add that some of the species of holly in North America, are, according to Asa Gray, in an exactly intermediate condition, or, as he expresses it, are more or less diœciously polygamous.

Let us now turn to the nectar-feeding insects; we may suppose the plant, of which we have been slowly increasing the nectar by continued selection, to be a common plant; and that certain insects depended in main part on its nectar for food. I could give many facts showing how anxious bees are to save time: for instance, their habit of cutting holes and sucking the nectar at the bases of certain flowers, which, with a very little more trouble, they can enter by the mouth. Bearing such facts in mind, it may be believed that under certain circumstances individual differences in the curvature or length of the proboscis, &c., too slight to be appreciated by us, might profit a bee or other insect, so that certain individuals would be able to obtain their food more quickly than others; and thus the communities to which they belonged would flourish and throw off many swarms inheriting the same peculiarities. The tubes of the corolla of the common red and incarnate clovers (Trifolium pratense and incarnatum) do not on a hasty glance appear to differ in length; yet the hive-bee can easily suck the nectar out of the incarnate clover, but not out of the common red clover, which is visited by humble-bees alone; so that whole fields of red clover offer in vain an abundant supply of precious nectar to the hive-bee. That this nectar is much liked by the hive-bee is certain; for I have repeatedly seen, but only in the autumn, many hive-bees sucking the flowers through holes bitten in the base of the tube by bumble-bees. The difference in the length of the corolla in the two kinds of clover, which determines the visits of the hive-bee, must be very trifling; for I have been assured that when red clover has been mown, the flowers of the second crop are somewhat smaller, and that these are visited by many hive-bees. I do not know whether this statement is accurate; nor whether another published statement can be trusted, namely, that the Ligurian bee which is generally considered a mere variety of the common hive-bee, and which freely crosses with it, is able to reach and suck the nectar of the red clover. Thus, in a country where this kind of clover abounded, it might be a great advantage to the hive-bee to have a slightly longer or differently constructed proboscis. On the other hand, as the fertility of this clover absolutely depends on bees visiting the flowers, if humble-bees were to become rare in any country, it might be a great advantage to the plant to have a shorter or more deeply divided corolla, so that the hive-bees should be enabled to suck its flowers. Thus I can understand how a flower and a bee might slowly become, either simultaneously or one after the other, modified and adapted to each other in the most perfect manner, by the continued preservation of all the individuals which presented slight deviations of structure mutually favourable to each other.

I am well aware that this doctrine of natural selection, exemplified in the above imaginary instances, is open to the same objections which were first urged against Sir Charles Lyell's noble views on "the modern changes of the earth, as illustrative of geology"; but we now seldom hear the agencies which we see still at work, spoken of as trifling or insignificant, when used in explaining the excavation of the deepest valleys of the formation of long lines of inland cliffs. Natural selection acts only by the preservation and accumulation of small inherited modifications, each profitable to the preserved being; and as modern geology has almost banished such views as the excavation of a great valley by a single diluvial wave, so will natural selection banish the belief of the continued creation of new organic beings, or of any great and sudden modification in their structure. * * *

CIRCUMSTANCES FAVOURABLE FOR THE PRODUCTION OF NEW FORMS THROUGH NATURAL SELECTION

This is an extremely intricate subject. A great amount of variability, under which term individual differences are always included, will evidently be favourable. A large number of individuals, by giving a better chance within any given period for the appearance of profitable variations, will conpensate for a lesser amount of variability in each individual, and is, I believe, a highly important element of success. Though Nature grants long periods of time for the work of natural selection, she does not grant an indefinite period; for as all organic beings are striving to seize on each place in the economy of nature, if any one species does not become modified and improved in a corresponding degree with its competitors, it will be exterminated. Unless favourable variations be inherited by some at least of the offspring, nothing can be effected by natural selection. The tendency to reversion may often check or prevent the work; but as this tendency has not prevented man from forming by selection numerous domestic races, why should it prevail against natural selection?

In the case of methodical selection, a breeder selects for some definite object, and if the individuals be allowed freely to intercross, his work will completely fail. But when many men, without intending to alter the breed, have a nearly common standard of perfection, and all try to procure and breed from the best animals, improvement surely but slowly follows from this unconscious process of selection, notwithstanding that there is no separation of selected individuals. Thus it will be under nature; for within a confined area, with some place in the natural polity not perfectly occupied, all the individuals varying in the right direction, though in different degrees, will tend to be preserved. But if the area be large, its several districts will almost certainly present different conditions of life; and then, if the same species undergoes modification in different districts, the newly-formed varieties will intercross on the confines of each. But we shall see in the sixth chapter that intermediate varieties, inhabiting intermediate districts, will in the long run generally be supplanted by one of the adjoining varieties. Intercrossing will chiefly affect those animals which unite for each birth and wander much, and which do not breed at a very quick rate. Hence with animals of this nature, for instance, birds, varieties will generally be confined to separated countries; and this I find to be the case. With hermaphrodite organisms which cross only occasionally, and likewise with animals which unite for each birth, but which wander little and can increase at a rapid rate, a new and improved variety might be quickly formed on any one spot, and might there maintain itself in a body and afterwards spread, so that the individuals of the new variety would chiefly cross together. On this principle, nurserymen always prefer saving seed from a large body of plants, as the chance of intercrossing is thus lessened.

Even with animals which unite for each birth, and which do not propagate rapidly, we must not assume that free intercrossing would always eliminate the effects of natural selection; for I can bring forward a considerable body of facts showing that within the same area, two varieties of the same animal may long remain distinct, from haunting different stations, from breeding at slightly different seasons, or from the individuals of each variety preferring to pair together.

Intercrossing plays a very important part in nature by keeping the individuals of the same species, or of the same variety, true and uniform in character. It will obviously thus act far more efficiently with those animals which unite for each birth; but, as already stated, we have reason to believe that occasional intercrosses take place with all animals and plants. Even if these take place only at long intervals of time, the young thus produced will gain so much in vigour and fertility over the offspring from long-continued self-fertilisation, that they will have a better chance of surviving and propagating their kind; and thus in the long run the influence of crosses, even at rare intervals, will be great. With respect to organic beings extremely low in the scale, which do not propagate sexually, nor conjugate, and which cannot possibly intercross, uniformity of character can be retained by them under the same conditions of life, only through the principle of inheritance, and through natural selection which will destroy any individuals departing from the proper type. If the conditions of life change and the form undergoes modification, uniformity of character can be given to the modified offspring, solely by natural selection preserving similar favourable variations.

Isolation, also, is an important element in the modification of species through natural selection. In a confined or isolated area, if not very large, the organic and inorganic conditions of life will generally be almost uniform; so that natural selection will tend to modify all the varying individuals of the same species in the same manner. Intercrossing with the inhabitants of the surrounding districts will, also, be thus prevented. Moritz Wagner has published an interesting essay on this subject, and has shown that the service rendered by isolation in preventing crosses between newly-formed varieties is probably greater even than I supposed. But from reasons already assigned I can by no means agree with this naturalist, that migration and isolation are necessary elements for the formation of new species. The importance of isolation is likewise great in preventing, after any physical change in the conditions, such as of climate, elevation of the land, &c., the immigration of better adapted organisms; and thus new places in the natural economy of the district will be left open to be filled up by the modification of the old inhabitants. Lastly, isolation will give time for a new variety to be improved at a slow rate; and this may sometimes be of much importance. If, however, an isolated area be very small, either from being surrounded by barriers, or from having very peculiar physical conditions, the total number of the inhabitants will be small; and this will retard the production of new species, through natural selection by decreasing the chances of favourable variations arising.

The mere lapse of time by itself does nothing, either for or against natural selection. I state this because it has been erroneously asserted that the element of time has been assumed by me to play an all-important part in modifying species, as if all the forms of life were necessarily undergoing change through some innate law. Lapse of time is only so far important, and its importance in this respect is great, that it gives a better chance of beneficial variations arising and of their being selected, accumulated, and fixed. It likewise tends to increase the direct action of the physical conditions of life, in relation to the constitution of each organism.

If we turn to nature to test the truth of these remarks, and look at any small isolated area, such as an oceanic island, although the number of species inhabiting it is small, as we shall see in our chapter on Geographical Distribution; yet of these species a very large proportion are endemic,—that is, have been produced there and nowhere else in the world. Hence an oceanic island at first sight seems to have been highly favourable for the production of new species. But we may thus deceive ourselves, for to ascertain whether a small isolated area, or a large open area like a continent, has been most favourable for the production of new organic forms, we ought to make the comparison within equal times; and this we are incapable of doing.

Although isolation is of great importance in the production of new species, on the whole I am inclined to believe that largeness of area is still more important, especially for the production of species which shall prove capable of enduring for a long period, and of spreading widely. Throughout a great and open area, not only will there be a better chance of favourable variations, arising from the large number of individuals of the same species there supported, but the conditions of life are much more complex from the large number of already existing species; and if some of these many species become modified and improved, others will have to be improved in a corresponding degree, or they will be exterminated. Each new form, also, as soon as it has been much improved, will be able to spread over the open and continuous area, and will thus come into competition with many other forms. Moreover, great areas, though now continuous, will often, owing to former oscillations of level, have existed in a broken condition; so that the good effects of isolation will generally, to a certain extent, have concurred. Finally, I conclude that, although small isolated areas have been in some respects highly favourable for the production of new species, yet that the course of modification will generally have been more rapid on large areas; and what is more important, that the new forms produced on large areas, which already have been victorious over many competitors, will be those that will spread most widely, and will give rise to the greatest number of new varieties and species. They will thus play a more important part in the changing history of the organic world.

In accordance with this view, we can, perhaps, understand some facts which will be again alluded to in our chapter on Geographical Distribution; for instance, the fact of the productions of the smaller continent of Australia now yielding before those of the larger Europæo-Asiatic area. Thus, also, it is that continental productions have everywhere become so largely naturalised on islands. On a small island, the race for life will have been less severe, and there will have been less modification and less extermination. Hence, we can understand how it is that the flora of Madeira, according to Oswald Heer, resembles to a certain extent the extinct tertiary flora of Europe. All fresh-water basins, taken together, make a small area compared with that of the sea or of the land. Consequently, the competition between fresh-water productions will have been less severe than elsewhere; new forms will have been then more slowly produced, and old forms more slowly exterminated. And it is in fresh-water basins that we find seven genera of Ganoid fishes, remnants of a once preponderant order: and in fresh water we find some of the most anomalous forms now known in the world as the Ornithorhynchus and Lepidosiren which, like fossils, connect to a certain extent orders at present widely sundered in the natural scale. These anomalous forms may be called living fossils; they have endured to the present day, from having inhabited a confined area, and from having been exposed to less varied, and therefore less severe, competition.

To sum up, as far as the extreme intricacy of the subject permits, the circumstances favourable and unfavourable for the production of new species through natural selection. I conclude that for terrestrial productions a large continental area, which has undergone many oscillations of level, will have been the most favourable for the production of many new forms of life, fitted to endure for a long time and to spread widely. Whilst the area existed as a continent, the inhabitants will have been numerous in individuals and kinds, and will have been subjected to severe competition. When converted by subsidence into large separate islands, there will still have existed many individuals of the same species on each island: intercrossing on the confines of the range of each new species will have been checked: after physical changes of any kind, immigration will have been prevented, so that new places in the polity of each island will have had to be filled up by the modification of the old inhabitants; and time will have been allowed for the varieties in each to become well modified and perfected. When, by renewed elevation, the islands were reconverted into a continental area, there will again have been very severe competition: the most favoured or improved varieties will have been enabled to spread: there will have been much extinction of the less improved forms, and the relative proportional numbers of the various inhabitants of the reunited continent will again have been changed; and again there will have been a fair field for natural selection to improve still further the inhabitants, and thus to produce new species.

That natural selection generally acts with extreme slowness I fully admit. It can act only when there are places in the natural polity of a district which can be better occupied by the modification of some of its existing inhabitants. The occurrence of such places will often depend on physical changes, which generally take place very slowly, and on the immigration of better adapted forms being prevented. As some few of the old inhabitants become modified, the mutual relations of others will often be disturbed; and this will create new places, ready to be filled up by better adapted forms, but all this will take place very slowly. Although all the individuals of the same species differ in some slight degree from each other, it would often be long before differences of the right nature in various parts of the organisation might occur. The result would often be greatly retarded by free intercrossing. Many will exclaim that these several causes are amply sufficient to neutralise the power of natural selection. I do not believe so. But I do believe that natural selection will generally act very slowly, only at long intervals of time, and only on a few of the inhabitants of the same region. I further believe that these slow, intermittent results accord well with what geology tells us of the rate and manner at which the inhabitants of the world have changed.

Slow though the process of selection may be, if feeble man can do much by artificial selection, I can see no limit to the amount of change, to the beauty and complexity of the coadaptations between all organic beings, one with another and with their physical conditions of life, which may have been effected in the long course of time through nature's power of selection, that is by the survival of the fittest. * * *

DIVERGENCE OF CHARACTER

The principle, which I have designated by this term, is of high importance, and explains, as I believe, several important facts. In the first place, varieties, even strongly-marked ones, though having somewhat of the character of species—as is shown by the hopeless doubts in many cases how to rank them—yet certainly differ far less from each other than do good and distinct species. Nevertheless, according to my view, varieties are species in the process of formation, or are, as I have called them, incipient species. How, then, does the lesser difference between varieties become augmented into the greater difference between species? That this does habitually happen, we must infer from most of the innumerable species throughout nature presenting well-marked differences; whereas varieties, the supposed prototypes and parents of future well-marked species, present slight and ill-defined differences. Mere chance, as we may call it, might cause one variety to differ in some character from its parents, and the offspring of this variety again to differ from its parent in the very same character and in a greater degree; but this alone would never account for so habitual and large a degree of difference as that between the species of the same genus.

As has always been my practice, I have sought light on this head from our domestic productions. We shall here find something analogous. It will be admitted that the production of races so different as short-horn and Hereford cattle, race and cart horses, the several breeds of pigeons, &c., could never have been effected by the mere chance accumulation of similar variations during many successive generations. In practice, a fancier is, for instance, struck by a pigeon having a slightly shorter beak; another fancier is struck by a pigeon having a rather longer beak; and on the acknowledged principle that "fanciers do not and will not admire a medium standard, but like extremes," they both go on (as has actually occurred with the sub-breeds of the tumbler-pigeon) choosing and breeding from birds with longer and longer beaks, or with shorter and shorter beaks. Again, we may suppose that at an early period of history, the men of one nation or district required swifter horses, whilst those of another required stronger and bulkier horses. The early differences would be very slight; but, in the course of time, from the continued selection of swifter horses in the one case, and of stronger ones in the other, the differences would become greater, and would be noted as forming two sub-breeds. Ultimately, after the lapse of centuries, these sub-breeds would become converted into two well-established and distinct breeds. As the differences became greater, the inferior animals with intermediate characters, being neither swift nor very strong, would not have been used for breeding, and will thus have tended to disappear. Here, then, we see in man's productions the action of what may be called the principle of divergence, causing differences, at first barely appreciable, steadily to increase, and the breeds to diverge in character, both from each other and from their common parent.

But how, it may be asked, can any analogous principle apply in nature? I believe it can and does apply most efficiently (though it was a long time before I saw how), from the simple circumstance that the more diversified the descendants from any one species become in structure, constitution, and habits, by so much will they be better enabled to seize on many and widely diversified places in the polity of nature, and so be enabled to increase in numbers.

We can clearly discern this in the case of animals with simple habits. Take the case of a carnivorous quadruped, of which the number that can be supported in any country has long ago arrived at its full average. If its natural power of increase be allowed to act, it can succeed in increasing (the country not undergoing any change in conditions) only by its varying descendants seizing on places at present occupied by other animals: some of them, for instance, being enabled to feed on new kinds of prey, either dead or alive; some inhabiting new stations, climbing trees, frequenting water, and some perhaps becoming less carnivorous. The more diversified in habits and structure the descendants of our carnivorous animals become, the more places they will be enabled to occupy. What applies to one animal will apply throughout all time to all animals—that is, if they vary—for otherwise natural selection can effect nothing. So it will be with plants. It has been experimentally proved, that if a plot of ground be sown with one species of grass, and a similar plot be sown with several distinct genera of grasses, a greater number of plants and a greater weight of dry herbage can be raised in the latter than in the former case. The same has been found to hold good when one variety and several mixed varieties of wheat have been sown on equal spaces of ground. Hence, if any one species of grass were to go on varying, and the varieties were continually selected which differed from each other in the same manner, though in a very slight degree, as do the distinct species and genera of grasses, a greater number of individual plants of this species, including its modified descendants, would succeed in living on the same piece of ground. And we know that each species and each variety of grass is annually sowing almost countless seeds; and is thus striving, as it may be said, to the utmost to increase in number. Consequently, in the course of many thousand generations, the most distinct varieties of any one species of grass would have the best chance of succeeding and of increasing in numbers, and thus of supplanting the less distinct varieties; and varieties, when rendered very distinct from each other, take the rank of species.

The truth of the principle that the greatest amount of life can be supported by great diversification of structure, is seen under many natural circumstances. In an extremely small area, especially if freely open to immigration, and where the contest between individual and individual must be very severe, we always find great diversity in its inhabitants. For instance, I found that a piece of turf, three feet by four in size, which had been exposed for many years to exactly the same conditions, supported twenty species of plants, and these belonged to eighteen genera and to eight orders, which shows how much these plants differed from each other. So it is with the plants and insects on small and uniform islets: also in small ponds of fresh water. Farmers find that they can raise most food by a rotation of plants belonging to the most different orders: nature follows what may be called a simultaneous rotation. Most of the animals and plants which live close round any small piece of ground, could live on it (supposing its nature not to be in any way peculiar), and may be said to be striving to the utmost to live there; but, it is seen, that where they come into the closest competition, the advantages of diversification of structure, with the accompanying differences of habit and constitution, determine that the inhabitants, which thus jostle each other most closely, shall, as a general rule, belong to what we call different genera and orders.

The same principle is seen in the naturalisation of plants through man's agency in foreign lands. It might have been expected that the plants which would succeed in becoming naturalised in any land would generally have been closely allied to the indigenes; for these are commonly looked at as specially created and adapted for their own country. It might also, perhaps, have been expected that naturalised plants would have belonged to a few groups more especially adapted to certain stations in their new homes. But the case is very different; and Alph. de Candolle has well remarked, in his great and admirable work, that floras gain by naturalisation, proportionally with the number of the native genera and species far more in new genera than in new species. To give a single instance: in the last edition of Dr. Asa Gray's 'Manual of the Flora of the Northern United States,' 260 naturalised plants are enumerated, and these belong to 162 genera. We thus see that these naturalised plants are of a highly diversified nature. They differ, moreover, to a large extent, from the indigenes, for out of the 162 naturalised genera, no less than 100 genera are not there indigenous, and thus a large proportional addition is made to the genera now living in the United States.

By considering the nature of the plants or animals which have in any country struggled successfully with the indigenes and have there become naturalised, we may gain some crude idea in what manner some of the natives would have to be modified, in order to gain an advantage over their compatriots; and we may at least infer that diversification of structure, amounting to new generic differences, would be profitable to them.

The advantage of diversification of structure in the inhabitants of the same region is, in fact, the same as that of the physiological division of labour in the organs of the same individual body—a subject so well elucidated by Milne Edwards. No physiologist doubts that a stomach adapted to digest vegetable matter alone, or flesh alone, draws most nutriment from these substances. So in the general economy of any land, the more widely and perfectly the animals and plants are diversified for different habits of life, so will a greater number of individuals be capable of there supporting themselves. A set of animals, with their organisation but little diversified, could hardly compete with a set more perfectly diversified in structure. It may be doubted, for instance, whether the Australian marsupials, which are divided into groups differing but little from each other, and feebly representing, as Mr. Waterhouse and others have remarked, our carnivorous, ruminant, and rodent mammals, could successfully compete with these well-developed orders. In the Australian mammals, we see the process of diversification in an early and incomplete stage of development.

THE PROBABLE EFFECTS OF THE ACTION OF NATURAL SELECTION THROUGH DIVERGENCE OF CHARACTER AND EXTINCTION, ON THE DESCENDANTS OF A COMMON ANCESTOR

After the foregoing discussion, which has been much compressed, we may assume that the modified descendants of any one species will succeed so much the better as they become more diversified in structure, and are thus enabled to encroach on places occupied by other beings. Now let us see how this principle of benefit being derived from divergence of character, combined with the principles of natural selection and of extinction, tends to act.

The accompanying diagram will aid us in understanding this rather perplexing subject. Let A to L represent the species of a genus large in its own country; these species are supposed to resemble each other in unequal degrees, as is so generally the case in nature, and as is represented in the diagram by the letters standing at unequal distances. I have said a large genus, because as we saw in the second chapter, on an average more species vary in large genera than in small genera; and the varying species of the large genera present a greater number of varieties. We have, also, seen that the species, which are the commonest and the most widely diffused, vary more than do the rare and restricted species. Let (A) be a common, widely-diffused, and varying species, belonging to a genus large in its own country. The branching and diverging dotted lines of unequal lengths proceeding from (A), may represent its varying offspring. The variations are supposed to be extremely slight, but

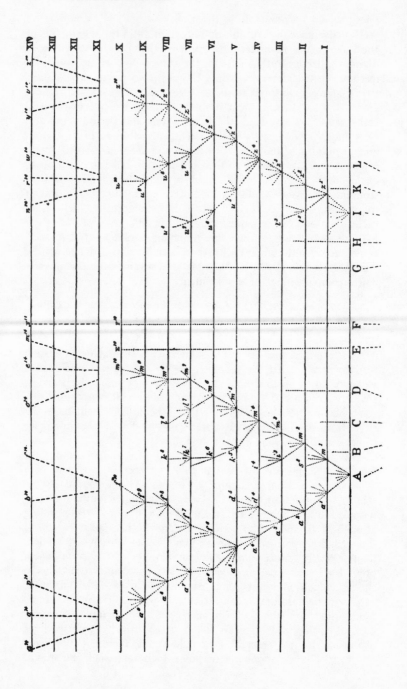

of the most diversified nature; they are not supposed all to appear simultaneously, but often after long intervals of time; nor are they all supposed to endure for equal periods. Only those variations which are in some way profitable will be preserved or naturally selected. And here the importance of the principle of benefit derived from divergence of character comes in; for this will generally lead to the most different or divergent variations (represented by the outer dotted lines) being preserved and accumulated by natural selection. When a dotted line reaches one of the horizontal lines, and is there marked by a small numbered letter, a sufficient amount of variation is supposed to have been accumulated to form it into a fairly well-marked variety, such as would be thought worthy of record in a systematic work.

The intervals between the horizontal lines in the diagram, may represent each a thousand or more generations. After a thousand generations, species (A) is supposed to have produced two fairly well-marked varieties, namely a^1 and m^1. These two varieties will generally still be exposed to the same conditions which made their parents variable, and the tendency to variability is in itself hereditary; consequently they will likewise tend to vary, and commonly in nearly the same manner as did their parents. Moreover, these two varieties, being only slightly modified forms, will tend to inherit those advantages which made their parent (A) more numerous than most of the other inhabitants of the same country; they will also partake of those more general advantages which made the genus to which the parent-species belonged, a large genus in its own country. And all these circumstances are favourable to the production of new varieties.

If, then, these two varieties be variable, the most divergent of their variations will generally be preserved during the next thousand generations. And after this interval, variety a^1 is supposed in the diagram to have produced variety a^2, which will, owing to the principle of divergence, differ more from (A) than did variety a^1. Variety m^1 is supposed to have produced two varieties, namely m^2 and s^2, differing from each other, and more considerably from their common parent (A). We may continue the process by similar steps for any length of time; some of the varieties, after each thousand generations, producing only a single variety, but in a more and more modified condition, some producing two or three varieties, and some failing to produce any. Thus the varieties or modified descendants of the common parent (A), will generally go on increasing in number and diverging in character. In the diagram the process is represented up to the ten-thousandth generation, and under a condensed and simplified form up to the fourteen-thousandth generation.

But I must here remark that I do not suppose that the process ever goes on so regularly as is represented in the diagram, though in itself made somewhat irregular, nor that it goes on continuously; it is far more probable that each form remains for long periods unaltered, and then again undergoes modification. Nor do I suppose that the most divergent varieties are invariably preserved: a medium form may often long endure, and may or may not produce more than one modified descendant; for natural selection will always act according to the nature of the places which are either unoccupied or not perfectly occupied by other beings; and this will depend on infinitely complex relations. But as a general rule, the more diversified in structure the descendants from any one species can be rendered, the more places they will be enabled to seize on, and the more their modified progeny will increase. In our diagram the line of succession is broken at regular intervals by small numbered letters marking the successive forms which have become sufficiently distinct to be recorded as varieties. But these breaks are imaginary, and might have been inserted anywhere, after intervals long enough to allow the accumulation of a considerable amount of divergent variation.

As all the modified descendants from a common and widely-diffused species, belonging to a large genus, will tend to partake of the same advantages which made their parent successful in life, they will generally go on multiplying in number as well as diverging in character: this is represented in the diagram by the several divergent branches proceeding from (A). The modified offspring from the later and more highly improved branches in the lines of descent, will, it is probable, often take the place of, and so destroy, the earlier and less improved branches: this is represented in the diagram by some of the lower branches not reaching to the upper horizontal lines. In some cases no doubt the process of modification will be confined to a single line of descent and the number of modified descendants will not be increased; although the amount of divergent modification may have been augmented. This case would be represented in the diagram, if all the lines proceeding from (A) were removed, excepting that from a^1 to a^{10}. In the same way the English racehorse and English pointer have apparently both gone on slowly diverging in character from their original stocks, without either having given off any fresh branches or races.

After ten thousand generations, species (A) is supposed to have produced three forms, a^{10}, f^{10}, and m^{10}, which, from having diverged in character during the successive generations, will have come to differ largely, but perhaps unequally, from each other and from their common parent. If we suppose the amount of change between each horizontal line in our diagram to be excessively small, these three forms may still be only well-marked varieties; but we have only to suppose the steps in the process of modification to be more numerous or greater in amount, to convert these three forms into doubtful or at least into well-defined species. Thus the diagram illustrates the steps by which the small differences distinguishing varieties are increased into the larger differences distinguishing species. By continuing the same process for a greater number of generations (as shown in the diagram in a condensed and simplified manner), we get eight species, marked by the letters between a^{14} and m^{14}, all descended from (A). Thus, as I believe, species are multiplied and genera are formed.

In a large genus it is probable that more than one species would vary. In the diagram I have assumed that a second species (I) has produced, by analogous steps, after ten thousand generations, either two well-marked varieties (w^{10} and z^{10}) or two species, according to the amount of change supposed to be represented between the horizontal lines. After fourteen thousand generations, six new species marked by the letters n^{14} to z^{14}, are supposed to have been produced. In any genus, the species which are already very different in character from each other, will generally tend to produce the greatest number of modified descendants; for these will have the best chance of seizing on new and widely different places in the polity of nature: hence in the diagram I have chosen the extreme species (A), and the nearly extreme species (I), as those which have largely varied, and have given rise to new varieties and species. The other nine species (marked by capital letters) of our original genus, may for long but unequal periods continue to transmit unaltered descendants; and this is shown in the diagram by the dotted lines unequally prolonged upwards.

But during the process of modification, represented in the diagram, another of our principles, namely that of extinction, will have played an important part. As in each fully stocked country natural selection necessarily acts by the selected form having some advantage in the struggle for life over other forms, there will be a constant tendency in the improved descendants of any one species to supplant and exterminate in each stage of descent their predecessors and their original progenitor. For it should be remembered that the competition will generally be most severe between those forms which are most nearly related to each other in habits, constitution, and structure. Hence all the intermediate forms between the earlier and later states, that is between the less and more improved states of the same species, as well as the original parent-species itself, will generally tend to become extinct. So it probably will be with many whole collateral lines of descent, which will be conquered by later and improved lines. If, however, the modified offspring of a species get into some distinct country, or become quickly adapted to some quite new station, in which offspring and progenitor do not come into competition, both may continue to exist.

If, then, our diagram be assumed to represent a considerable amount of modification, species (A) and all the earlier varieties will have become extinct, being replaced by eight new species (a^{14} to m^{14}); and species (I) will be replaced by six (n^{14} to z^{14}) new species.

But we may go further than this. The original species of our genus were supposed to resemble each other in unequal degrees, as is so generally the case in nature; species (A) being more nearly related to B. C, and D, than to the other species; and species (I) more to G, H, K, L, than to the others. These two species (A) and (I) were also supposed to be very common and widely diffused species, so that they must originally have had some advantage over most of the other species of the genus. Their modified descendants, fourteen in number at the fourteen-thousandth generation, will probably have inherited some of the same advantages: they have also been modified and improved in a diversified manner at each stage of descent, so as to have become adapted to many related places in the natural economy of their country. It seems, therefore, extremely probable that they will have taken the places of, and thus exterminated not only their parents (A) and (I), but likewise some of the original species which were most nearly related to their parents. Hence very few of the original species will have transmitted offspring to the fourteen-thousandth generation. We may suppose that only one, (F), of the two species (E and F) which were least closely related to the other nine original species, has transmitted descendants to this late stage of descent.

The new species in our diagram descended from the original eleven species, will now be fifteen in number. Owing to the divergent tendency of natural selection, the extreme amount of difference in character between species a^{14} and z^{14} will be much greater than that between the most distinct of the original eleven species. The new species, moreover, will be allied to each other in a widely different manner. Of the eight descendants from (A) the three marked a^{14}, q^{14}, p^{14}, will be nearly related from having recently branched off from a^{10}, b^{14}, and f^{14}, from having diverged at an earlier period from a^5, will be in some degree distinct from the three first-named species; and lastly, o^{14}, e^{14}, and m^{14}, will be nearly related one to the other, but, from having diverged at the first commencement of the process of modification, will be widely different from the other five species, and may constitute a sub-genus or a distinct genus.

The six descendants from (I) will form two sub-genera or genera. But as the original species (I) differed largely from (A), standing nearly at the extreme end of the original genus, the six descendants from (I) will, owing to inheritance alone, differ considerably from the eight descendants from (A); the two groups, moreover, are supposed to have gone on diverging in different directions. The intermediate species, also (and this is a very important consideration), which connected the original species (A) and (I), have become, excepting (F), extinct, and have left no descendants. Hence the six new species descended from (I), and the eight descendants from (A), will have to be ranked as very distinct genera, or even as distinct sub-families.

Thus it is, as I believe, that two or more genera are produced by descent with modification, from two or more species of the same genus. And the two or more parent-species are supposed to be descended from some one species of an earlier genus. In our diagram, this is indicated by the broken lines, beneath the capital letters, converging in sub-branches downwards towards a single point; this point represents a species, the supposed progenitor of our several new sub-genera and genera.

It is worth while to reflect for a moment on the character of the new species F^{14}, which is supposed not to have diverged much in character, but to have retained the form of (F), either unaltered or altered only in a slight degree. In this case, its affinities to the other fourteen new species will be of a curious and circuitous nature. Being descended from a form which stood between the parent-species (A) and (I), now supposed to be extinct and unknown, it will be in some degree intermediate in character between the two groups descended from these two species. But as these two groups have gone on diverging in character from the type of their parents, the new species (F^{14}) will not be directly intermediate between them, but rather between types of the two groups; and every naturalist will be able to call such cases before his mind.

In the diagram, each horizontal line has hitherto been supposed to represent a thousand generations, but each may represent a million or more generations; it may also represent a section of the successive strata of the earth's crust including extinct remains. We shall, when we come to our chapter on Geology, have to refer again to this subject, and I think we shall then see that the diagram throws light on the affinities of extinct beings, which, though generally belonging to the same orders, families, or genera, with those now living, yet are often, in some degree, intermediate in character between existing groups; and we can understand this fact, for the extinct species lived at various remote epochs when the branching lines of descent had diverged less.

I see no reason to limit the process of modification, as now explained, to the formation of genera alone. If, in the diagram, we suppose the amount of change, represented by each successive group of diverging dotted lines to be great, the forms marked a^{14} to p^{14}, those marked b^{14} and f^{14}, and those marked o^{14} to m^{14}, will form three very distinct genera. We shall also have two very distinct genera descended from(I), differing widely from the descendants of (A). These two groups of genera will thus form two distinct families, or orders, according to the amount of divergent modification supposed to be represented in the diagram. And the two new families, or orders, are descended from two species of the original genus, and these are supposed to be descended from some still more ancient and unknown form.

We have seen that in each country it is the species belonging to the larger genera which oftenest present varieties or incipient species. This, indeed, might have been expected; for, as natural selection acts through one form having some advantage over other forms in the struggle for existence, it will chiefly act on those which already have some advantage; and the largeness of any group shows that its species have inherited from a common ancestor some advantage in common. Hence, the struggle for the production of new and modified descendants will mainly lie between the larger groups which are all trying to increase in number. One large group will slowly conquer another large group, reduce its numbers, and thus lessen its chance of further variation and improvement. Within the same large group, the later and more highly perfected subgroups, from branching out and seizing on many new places in the polity of Nature, will constantly tend to supplant and destroy the earlier and less improved sub-groups. Small and broken groups and sub-groups will finally disappear. Looking to the future, we can predict that the groups of organic beings which are now large and triumphant, and which are least broken up, that is, which have as yet suffered least extinction, will, for a long period, continue to increase. But which groups will ultimately prevail, no man can predict; for we know that many groups formerly most extensively developed, have now become extinct. Looking still more remotely to the future, we may predict that, owing to the continued and steady increase of the larger groups, a multitude of smaller groups will become utterly extinct, and leave no modified descendants; and consequently that, of the species living at any one period, extremely few will transmit descendants to a remote futurity. I shall have to return to this subject in the chapter on Classification, but I may add that as, according to this view, extremely few of the more ancient species have transmitted descendants to the present day, and, as all the descendants of the same species form a class, we can understand how it is that there exist so few classes in each main division of the animal and vegetable kingdoms. Although few of the most ancient species have left modified descendants, yet, at remote geological periods, the earth may have been almost as well peopled with species of many genera, families, orders, and classes, as at the present time.

ON THE DEGREE TO WHICH ORGANISATION TENDS TO ADVANCE

Natural Selection acts exclusively by the preservation and accumulation of variations, which are beneficial under the organic and inorganic conditions to which each creature is exposed at all periods of life. The ultimate result is that each creature tends to become more and more improved in relation to its conditions. This improvement inevitably leads to the gradual advancement of the organisation of the greater number of living beings throughout the world. But here we enter on a very intricate subject, for naturalists have not defined to each other's satisfaction what is meant by an advance in organisation. Amongst the vertebrata the degree of intellect and an approach in structure to man clearly come into play. It might be thought that the amount of change which the various parts and organs pass through in their development from the embryo to maturity would suffice as a standard of comparison; but there are cases, as with certain parasitic crustaceans, in which several parts of the structure become less perfect, so that the mature animal cannot be called higher than its larva. Von Baer's standard seems the most widely applicable and the best, namely, the amount of differentiation of the parts of the same organic being, in the adult state as I should be inclined to add, and their specialisation for different functions; or, as Milne Edwards would express it, the completeness of the division is physiological labour. But we shall see how obscure this subject is if we look, for instance, to fishes, amongst which some naturalists rank those as highest which, like the sharks, approach nearest to amphibians; whilst other naturalists rank the common bony or teleostean fishes as the highest, inasmuch as they are most strictly fish-like and differ most from the other vertebrate classes. We see still more plainly the obscurity of the subject by turning to plants, amongst which the standard of intellect is of course quite excluded; and here some botanists rank those plants as highest which have every organ, as sepals, petals, stamens, and pistils, fully developed in each flower; whereas other botanists, probably with more truth, look at the plants which have their several organs much modified and reduced in number as the highest.

If we take as the standard of high organisation, the amount of differentiation and specialisation of the several organs in each being when adult (and this will include the advancement of the brain for intellectual purposes), natural selection clearly leads towards this standard: for all physiologists admit that the specialisation of organs, inasmuch as in this state they perform their functions better, is an advantage to each being; and hence the accumulation of variations tending towards specialisation is within the scope of natural selection. On the other hand, we can see, bearing in mind that all organic beings are striving to increase at a high ratio and to seize on every unoccupied or less well occupied place in the economy of nature, that it is quite possible for natural selection gradually to fit a being to a situation in which several organs would be superfluous or useless: in such cases there would be retrogression in the scale of organisation. Whether organisation on the whole has actually advanced from the remotest geological periods to the present day will be more conveniently discussed in our chapter on Geological Succession.

But it may be objected that if all organic beings thus tend to rise in the scale, how is it that throughout the world a multitude of the lowest forms still exist; and how is it that in each great class some forms are far more highly developed than others? Why have not the more highly developed forms everywhere supplanted and exterminated the lower? Lamarck, who believed in an innate and inevitable tendency towards perfection in all organic beings, seems to have felt this difficulty so strongly, that he was led to suppose that new and simple forms are continually being produced by spontaneous generation. Science has not as yet proved the truth of this belief, whatever the future may reveal. On our theory the continued existence of lowly organisms offers no difficulty; for natural selection, or the survival of the fittest, does not necessarily include progressive development—it only takes advantage of such variations as arise and are beneficial to each creature under its complex relations of life. And it may be asked what advantage, as far as we can see, would it be to an infusorian animalcule—to an intestinal worm—or even to an earthworm, to be highly organised. If it were no advantage, these forms would be left, by natural selection, unimproved or but little improved, and might remain for indefinite ages in their present lowly condition. And geology tells us that some of the lowest forms, as the infusoria and rhizopods, have remained for an enormous period in nearly their present state. But to suppose that most of the many now existing low forms have not in the least advanced since the first dawn of life would be extremely rash; for every naturalist who has dissected some of the beings now ranked as very low in the scale, must have been struck with their really wondrous and beautiful organisation.

Nearly the same remarks are applicable if we look to the different grades of organisation within the same great group; for instance, in the vertebrata, to the co-existence of mammals and fish —amongst mammalia, to the co-existence of man and the ornithorhynchus—amongst fishes, to the co-existence of the shark and the lancelet (Amphioxus), which latter fish in the extreme simplicity of its structure approaches the invertebrate classes. But mammals and fish hardly come into competition with each other; the advancement of the whole class of mammals, or of certain members in this class, to the highest grade would not lead to their taking the place of fishes. Physiologists believe that the brain must be bathed by warm blood to be highly active, and this requires aërial respiration; so that warm-blooded mammals when inhabiting the water lie under a disadvantage in having to come continually to the surface to breathe. With fishes, members of the shark family would not tend to supplant the lancelet; for the lancelet, as I hear from Fritz Müller, has as sole companion and competitor on the barren sandy shore of South Brazil, an anomalous annelid. The three lowest orders of mammals, namely, marsupials, edentata, and rodents, co-exist in South America in the same region with numerous monkeys, and probably interfere little with each other. Although organisation, on the whole, may have advanced and be still advancing throughout the world, yet the scale will always present many degrees of perfection; for the high advancement of certain whole classes, or of certain members of each class, does not at all necessarily lead to the extinction of those groups with which they do not enter into close competition. In some cases, as we shall hereafter see, lowly organised forms appear to have been preserved to the present day, from inhabiting confined or peculiar stations, where they have been subjected to less severe competition, and where their scanty numbers have retarded the chance of favourable variations arising.

Finally, I believe that many lowly organised forms now exist throughout the world, from various causes. In some cases variations or individual differences of a favourable nature may never have arisen for natural selection to act on and accumulate. In no case, probably, has time sufficed for the utmost possible amount of development. In some few cases there has been what we must call retrogression of organisation. But the main cause lies in the fact that under very simple conditions of life a high organisation would be of no service,—possibly would be of actual disservice, as being of a more delicate nature, and more liable to be put out of order and injured.

Looking to the first dawn of life, when all organic beings, as we may believe, presented the simplest structure, how, it has been asked, could the first steps in the advancement or differentiation of parts have arisen? Mr. Herbert Spencer would probably answer that, as soon as simple unicellular organism came by growth or division to be compounded of several cells, or became attached to any supporting surface, his law "that homologous units of any order become differentiated in proportion as their relations to incident forces become different" would come into action. But as we have no facts to guide us, speculation on the subject is almost useless. It is, however, an error to suppose that there would be no struggle for existence, and, consequently, no natural selection, until many forms had been produced: variations in a single species inhabiting an isolated station might be beneficial, and thus the whole mass of individuals might be modified, or two distinct forms might arise. But, as I remarked towards the close of the Introduction, no one ought to feel surprise at much remaining as yet unexplained on the origin of species, if we make due allowance for our profound ignorance on the mutual relations of the inhabitants of the world at the present time, and still more so during past ages. * * *

SUMMARY OF CHAPTER

If under changing conditions of life organic beings present individual differences in almost every part of their structure, and this cannot be disputed; if there be, owing to their geometrical rate of increase, a severe struggle for life at some age, season, or year, and this certainly cannot be disputed; then, considering the infinite complexity of the relations of all organic beings to each other and to their conditions of life, causing an infinite diversity in structure, constitution, and habits, to be advantageous to them, it would be a most extraordinary fact if no variations had ever occurred useful to each being's own welfare, in the same manner as so many variations have occurred useful to man. But if variations useful to any organic being ever do occur, assuredly individuals thus characterised will have the best chance of being preserved in the struggle for life; and from the strong principle of inheritance, these will tend to produce offspring similarly characterised. This principle of preservation, or the survival of the fittest, I have called Natural Selection. It leads to the improvement of each creature in relation to its organic and inorganic conditions of life; and consequently, in most cases, to what must be regarded as an advance in organisation. Nevertheless, low and simple forms will long endure if well fitted for their simple conditions of life.

Natural selection, on the principle of qualities being inherited at corresponding ages, can modify the egg, seed, or young, as easily as the adult. Amongst many animals, sexual selection will have given its aid to ordinary selection, by assuring to the most vigorous and best adapted males the greatest number of offspring. Sexual selection will also give characters useful to the males alone, in their struggles or rivalry with other males; and these characters will be transmitted to one sex or to both sexes, according to the form of inheritance which prevails.

Whether natural selection has really thus acted in adapting the various forms of life to their several conditions and stations, must be judged by the general tenor and balance of evidence given in the following chapters. But we have already seen how it entails extinction; and how largely extinction has acted in the world's history, geology plainly declares. Natural selection, also leads to divergence of character; for the more organic beings diverge in structure, habits, and constitution, by so much the more can a large number be supported on the area,—of which we see proof by looking to the inhabitants of any small spot, and to the productions naturalised in foreign lands. Therefore, during the modification of the descendants of any one species, and during the incessant struggle of all species to increase in numbers, the more diversified the descendants become, the better will be their chance of success in the battle for life. Thus the small differences distinguishing varieties of the same species, steadily tend to increase, till they equal the greater differences between species of the same genus, or even of distinct genera.

We have seen that it is the common, the widely-diffused and widely-ranging species, belonging to the larger genera within each class, which vary most; and these tend to transmit to their modified offspring that superiority which now makes them dominant in their own countries. Natural selection, as has just been remarked, leads to divergence of character and to much extinction of the less improved and intermediate forms of life. On these principles, the nature of the affinities, and the generally well-defined distinctions between the innumerable organic beings in each class throughout the world, may be explained. It is a truly wonderful fact—the wonder of which we are apt to overlook from familiarity—that all animals and all plants throughout all time and space should be related to each other in groups, subordinate to groups, in the manner which we everywhere behold—namely, varieties of the same species most closely related, species of the same genus less closely and unequally related, forming sections and sub-genera, species of distinct genera much less closely related, and genera related in different degrees, forming sub-families, families, orders, sub-classes and classes. The several subordinate groups in any class cannot be ranked in a single file, but seem clustered round points, and these round other points, and so on in almost endless cycles. If species had been independently created, no explanation would have been possible of this kind of classification; but it is explained through inheritance and the complex action of natural selection, entailing extinction and divergence of character, as we have seen illustrated in the diagram.

The affinities of all the beings of the same class have sometimes been represented by a great tree. I believe this simile largely speaks the truth. The green and budding twigs may represent existing species; and those produced during former years may represent the long succession of extinct species. At each period of growth all the growing twigs have tried to branch out on all sides, and to overtop and kill the surrounding twigs and branches, in the same manner as species and groups of species have at all times overmastered other species in the great battle for life. The limbs divided into great branches, and these into lesser and lesser branches, were themselves once, when the tree was young, budding twigs, and this connection of the former and present buds by ramifying branches may well represent the classification of all extinct and living species in groups subordinate to groups. Of the many twigs which flourished when the tree was a mere bush, only two or three, now grown into great branches, yet survive and bear the other branches; so with the species which lived during long-past geological periods, very few have left living and modified descendants. From the first growth of the tree, many a limb and branch has decayed and dropped off; and these fallen branches of various sizes may represent those whole orders, families, and genera which have now no living representatives, and which are known to us only in a fossil state. As we here and there see a thin straggling branch springing from a fork low down in a tree, and which by some chance has been favoured and is still alive on its summit, so we occasionally see an animal like the Ornithorhynchus or Lepidosiren, which in some small degree connects by its affinities two large branches of life, and which has apparently been saved from fatal competition by having inhabited a protected station. As buds give rise by growth to fresh buds, and these, if vigorous, branch out and overtop on all sides many a feebler branch, so by generation I believe it has been with the great Tree of Life, which fills with its dead and broken branches the crust of the earth, and covers the surface with its ever-branching and beautiful ramifications.

Chapter VI

DIFFICULTIES OF THE THEORY

Difficulties of the theory of descent with modification—Absence or rarity of transitional varieties—Transitions in habits of life—Diversified habits in the same species—Species with habits widely differ-

ent *from those of their allies—Organs of extreme perfection— Modes of transition—Cases of difficulty—Natura non facit saltum —Organs of small importance—Organs not in all cases absolutely perfect—The law of unity of type and of the conditions of existence embraced by the theory of natural selection.*

Long before the reader has arrived at this part of my work, a crowd of difficulties will have occurred to him. Some of them are so serious that to this day I can hardly reflect on them without being in some degree staggered; but, to the best of my judgment, the greater number are only apparent, and those that are real are not, I think, fatal to the theory.

These difficulties and objections may be classed under the following heads:—First, why, if species have descended from other species by fine gradations, do we not everywhere see innumerable transitional forms? Why is not all nature in confusion instead of the species being, as we see them, well defined?

Secondly, is it possible that an animal having, for instance, the structure and habits of a bat, could have been formed by the modification of some other animal with widely different habits and structure? Can we believe that natural selection could produce, on the one hand, an organ of trifling importance, such as the tail of a giraffe, which serves as a fly-flapper, and, on the other hand, an organ so wonderful as the eye?

Thirdly, can instincts be acquired and modified through natural selection? What shall we say to the instinct which leads the bee to make cells, and which has practically anticipated the discoveries of profound mathematicians?

Fourthly, how can we account for species, when crossed, being sterile and producing sterile offspring, whereas, when varieties are crossed, their fertility is unimpaired?

The two first heads will here be discussed; some miscellaneous objections in the following chapter; Instinct and Hybridism in the two succeeding chapters.

ON THE ABSENCE OR RARITY OF TRANSITIONAL VARIETIES

As natural selection acts solely by the preservation of profitable modifications, each new form will tend in a fully-stocked country to take the place of, and finally to exterminate, its own less improved parent-form and other less favoured forms with which it comes into competition. Thus extinction and natural selection go hand in hand. Hence, if we look at each species as descended from some unknown form, both the parent and all the transitional varieties will generally have been exterminated by the very process of the formation and perfection of the new form.

But, as by this theory innumerable transitional forms must have existed, why do we not find them embedded in countless numbers in the crust of the earth? It will be more convenient to discuss this question in the chapter on the Imperfection of the Geological Record; and I will here only state that I believe the answer mainly lies in the record being incomparably less perfect than is generally supposed. The crust of the earth is a vast museum; but the natural collections have been imperfectly made, and only at long intervals of time.

But it may be urged that when several closely-allied species inhabit the same territory, we surely ought to find at the present time many transitional forms. Let us take a simple case: in travelling from north to south over a continent, we generally meet at successive intervals with closely allied or representative species, evidently filling nearly the same place in the natural economy of the land. These representative species often meet and interlock; and as the one becomes rarer and rarer, the other becomes more and more frequent, till the one replaces the other. But if we compare these species where they intermingle, they are generally as absolutely distinct from each other in every detail of structure as are specimens taken from the metropolis inhabited by each. By my theory these allied species are descended from a common parent; and during the process of modification, each has become adapted to the conditions of life of its own region, and has supplanted and exterminated its original parent-form and all the transitional varieties between its past and present states. Hence we ought not to expect at the present time to meet with numerous transitional varieties in each region, though they must have existed there, and may be embedded there in a fossil condition. But in the intermediate region, having intermediate conditions of life, why do we not now find closely-linking intermediate varieties? This difficulty for a long time quite confounded me. But I think it can be in large part explained.

In the first place we should be extremely cautious in inferring, because an area is now continuous, that it has been continuous during a long period. Geology would lead us to believe that most continents have been broken up into islands even during the later tertiary periods; and in such islands distinct species might have been separately formed without the possibility of intermediate varieties existing in the intermediate zones. By changes in the form of the land and of climate, marine areas now continuous must often have existed within recent times in a far less continuous and uniform condition than at present. But I will pass over this way of escaping from the difficulty; for I believe that many perfectly defined species have been formed on strictly continuous areas; though I do not doubt that the formerly broken condition of areas now continuous, has played an important part in the formation of new species, more especially with freely-crossing and wandering animals.

In looking at species as they are now distributed over a wide area, we generally find them tolerably numerous over a large territory, then becoming somewhat abruptly rarer and rarer on the confines, and finally disappearing. Hence the neutral territory, between two representative species is generally narrow in comparison with the territory proper to each. We see the same fact in ascending mountains, and sometimes it is quite remarkable how abruptly, as Alph. de Candolle has observed, a common alpine species disappears. The same fact has been noticed by E. Forbes in sounding the depths of the sea with the dredge. To those who look at climate and the physical conditions of life as the all-important elements of distribution, these facts ought to cause surprise, as climate and height or depth graduate away insensibly. But when we bear in mind that almost every species, even in its metropolis, would increase immensely in numbers, were it not for other competing species; that nearly all either prey on or serve as prey for others; in short, that each organic being is either directly or indirectly related in the most important manner to other organic beings,—we see that the range of the inhabitants of any country by no means exclusively depends on insensibly changing physical conditions, but in a large part on the presence of other species, on which it lives, or by which it is destroyed, or with which it comes into competition; and as these species are already defined objects, not blending one into another by insensible gradations, the range of any one species, depending as it does on the range of others, will tend to be sharply defined. Moreover, each species on the confines of its range, where it exists in lessened numbers, will, during fluctuations in the number of its enemies or of its prey, or in the nature of the seasons, be extremely liable to utter extermination; and thus its geographical range will come to be still more sharply defined.

As allied or representative species, when inhabiting a continuous area, are generally distributed in such a manner that each has a wide range, with a comparatively narrow neutral territory between them, in which they become rather suddenly rarer and rarer; then, as varieties do not essentially differ from species, the same rule will probably apply to both; and if we take a varying species inhabiting a very large area, we shall have to adapt two varieties to two large areas, and a third variety to a narrow intermediate zone. The intermediate variety, consequently, will exist in lesser numbers from inhabiting a narrow and lesser area; and practically, as far as I can make out, this rule holds good with varieties in a state of nature. I have met with striking instances of the rule in the case of varieties intermediate between well-marked varieties in the genus Balanus. And it would appear from information given me by Mr. Watson, Dr. Asa Gray, and Mr. Wollaston, that generally, when varieties intermediate between two other forms occur, they are much rarer numerically than the forms which they connect. Now, if we may trust these facts and inferences, and conclude that varieties linking two other varieties together generally have existed in lesser numbers than the forms which they connect, then we can understand why intermediate varieties should not endure for very long periods: —why, as a general rule, they should be exterminated and disappear, sooner than the forms which they originally linked together.

For any form existing in lesser numbers would, as already remarked, run a greater chance of being exterminated than one existing in large numbers; and in this particular case the intermediate form would be eminently liable to the inroads of closely-allied forms existing on both sides of it. But it is a far more important consideration, that during the process of further modification, by which two varieties are supposed to be converted and perfected into two distinct species, the two which exist in larger numbers, from inhabiting larger areas, will have a great advantage over the intermediate variety, which exists in smaller numbers in a narrow and intermediate zone. For forms existing in larger numbers will have a better chance, within any given period, of presenting further favourable variations for natural selection to seize on, than will the rarer forms which exist in lesser numbers. Hence, the more common forms, in the race for life, will tend to beat and supplant the less common forms, for these will be more slowly modified and improved. It is the same principle which, as I believe, accounts for the common species in each country, as shown in the second chapter, presenting on an average a greater number of well-marked varieties than do the rarer species. I may illustrate what I mean by supposing three varieties of sheep to be kept, one adapted to an extensive mountainous region; a second to a comparatively narrow, hilly tract; and a third to the wide plains at the base; and that the inhabitants are all trying with equal steadiness and skill to improve their stocks by selection; the chances in this case will be strongly in favour of the great holders on the mountains or on the plains, improving their breeds more quickly than the small holders on the intermediate narrow, hilly tract; and consequently the improved mountain or plain breed will soon take the place of the less improved hill breed; and thus the two breeds, which originally existed in greater numbers, will come into close contact with each other, without the interposition of the supplanted, intermediate hill variety.

To sum up, I believe that species come to be tolerably well-defined objects, and do not at any one period present an inextricable chaos of varying and intermediate links; first, because new varieties are very slowly formed, for variation is a slow process, and natural selection can do nothing until favourable individual differences or variations occur, and until a place in the natural polity of the country can be better filled by some modification of some one or more of its inhabitants. And such new places will depend on slow changes of climate, or on the occasional immigration of new inhabitants, and, probably, in a still more important degree, on some of the old inhabitants becoming slowly modified, with the new forms thus produced, and the old ones acting and reacting on each other. So that, in any one region and at any one time, we ought to see only a few species presenting slight modifications of structure in some degree permanent; and this assuredly we do see.

Secondly, areas now continuous must often have existed within the recent period as isolated portions, in which many forms, more especially amongst the classes which unite for each birth and wander much, may have separately been rendered sufficiently distinct to rank as representative species. In this case, intermediate varieties between the several representative species and their common parent, must formerly have existed within each isolated portion of the land, but these links during the process of natural selection will have been supplanted and exterminated, so that they will no longer be found in a living state.

Thirdly, when two or more varieties have been formed in different portions of a strictly continuous area, intermediate varieties will, it is probable, at first have been formed in the intermediate zones, but they will generally have had a short duration. For these intermediate varieties will, from reasons already assigned (namely from what we know of the actual distribution of closely allied or representative species, and likewise of acknowledged varieties), exist in the intermediate zones in lesser numbers than the varieties which they tend to connect. From this cause alone the intermediate varieties will be liable to accidental extermination; and during the process of further modification through natural selection, they will almost certainly be beaten and supplanted by the forms which they connect; for these from existing in greater numbers will, in the aggregate, present more varieties, and thus be further improved through natural selection and gain further advantages.

Lastly, looking not to any one time, but to all time, if my theory be true, numberless intermediate varieties, linking closely together all the species of the same group, must assuredly have existed; but the very process of natural selection constantly tends, as has been so often remarked, to exterminate the parent-forms and the intermediate links. Consequently evidence of their former existence could be found only amongst fossil remains, which are preserved, as we shall attempt to show in a future chapter, in an extremely imperfect and intermittent record.

ON THE ORIGIN AND TRANSITIONS OF ORGANIC BEINGS WITH PECULIAR HABITS AND STRUCTURE

It has been asked by the opponents of such views as I hold, how, for instance, could a land carnivorous animal have been converted into one with aquatic habits; for how could the animal in its transitional state have subsisted? It would be easy to show that there now exist carnivorous animals presenting close intermediate grades from strictly terrestrial to aquatic habits; and as each exists by a struggle for life, it is clear that each must be well adapted to its place in nature. Look at the Mustela vison of North America, which has webbed feet, and which resembles an otter in its fur, short legs, and form of tail. During the summer this animal dives for and preys on fish, but during the long winter it leaves the frozen waters, and preys, like other pole-cats, on mice and land animals. If a different case had been taken, and it had been asked how an insectivorous quadruped could possibly have been converted into a flying bat, the question would have been far more difficult to answer. Yet I think such difficulties have little weight.

Here, as on other occasions, I lie under a heavy disadvantage, for, out of the many striking cases which I have collected, I can only give one or two instances of transitional habits and structures in allied species; and of diversified habits, either constant or occasional, in the same species. And it seems to me that nothing less than a long list of such cases is sufficient to lessen the difficulty in any particular case like that of the bat.

Look at the family of squirrels; here we have the finest gradation from animals with their tails only slightly flattened, and from others, as Sir J. Richardson has remarked, with the posterior part of their bodies rather wide and with the skin on their flanks rather full, to the so-called flying squirrels; and flying squirrels have their limbs and even the base of the tail united by a broad expanse of skin, which serves as a parachute and allows them to glide through the air to an astonishing distance from tree to tree. We cannot doubt that each structure is of use to each kind of squirrel in its own country, by enabling it to escape birds or beasts of prey, to collect food more quickly, or, as there is reason to believe, to lessen the danger from occasional falls. But it does not follow from this fact that the structure of each squirrel is the best that it is possible to conceive under all possible conditions. Let the climate and vegetation change, let other competing rodents or new beasts of prey immigrate, or old ones become modified, and all analogy would lead us to believe that some at least of the squirrels would decrease in numbers or become exterminated, unless they also become modified and improved in structure in a corresponding manner. Therefore, I can see no difficulty, more especially under changing conditions of life, in the continued preservation of individuals with fuller and fuller flank-membranes, each modification being useful, each being propagated, until by the accumulated effects of this process of natural selection, a perfect so-called flying squirrel was produced.

Now look at the Galeopithecus or so-called flying lemur, which formerly was ranked amongst bats, but is now believed to belong to the Insectivora. An extremely wide flank-membrane stretches from the corners of the jaw to the tail, and includes the limbs with the elongated fingers. This flank-membrane is furnished with an extensor muscle. Although no graduated links of structure, fitted for gliding through the air, now connect the Galeopithecus with the other Insectivora, yet there is no difficulty in supposing that such links formerly existed, and that each was developed in the same manner as with the less perfectly gliding squirrels; each grade of structure having been useful to its possessor. Nor can I see any insuperable difficulty in further believing that the membrane connected fingers and fore-arm of the Galeopithecus might have been greatly lengthened by natural selection; and this, as far as the organs of flight are concerned, would have converted the animal into a bat. In certain bats in which the wing-membrane extends from the top of the shoulder to the tail and includes the hind-legs, we perhaps see traces of an apparatus originally fitted for gliding through the air rather than for flight.

If about a dozen genera of birds were to become extinct, who would have ventured to surmise that birds might have existed which used their wings solely as flappers, like the logger-headed duck (Micropterus of Eyton); as fins in the water and as front-legs on the land, like the penguin; as sails, like the ostrich; and functionally for no purpose, like the Apteryx? Yet the structure of each of these birds is good for it, under the conditions of life to which it is exposed, for each has to live by a struggle; but it is not necessarily the best possible under all possible conditions. It must not be inferred from these remarks that any of the grades of wing-structure here alluded to, which perhaps may all be the result of disuse, indicate the steps by which birds actually acquired their perfect power of flight; but they serve to show what diversified means of transition are at least possible.

Seeing that a few members of such water-breathing classes as the Crustacea and Mollusca are adapted to live on the land; and seeing that we have flying birds and mammals, flying insects of the most diversified types, and formerly had flying reptiles, it is conceivable that flying-fish, which now glide far through the air, slightly rising and turning by the aid of their fluttering fins, might have been modified into perfectly winged animals. If this had been effected, who would have ever imagined that in an early transitional state they had been the inhabitants of the open ocean, and had used their incipient organs of flight exclusively, as far as we know, to escape being devoured by other fish?

When we see any structure highly perfected for any particular habit, as the wings of a bird for flight, we should bear in mind that animals displaying early transitional grades of the structure will seldom have survived to the present day, for they will have been supplanted by their successors, which were gradually rendered more perfect through natural selection. Furthermore, we may conclude that transitional states between structures fitted for very different habits of life will rarely have been developed at an early period in great numbers and under many subordinate forms. Thus, to return to our imaginary illustration of the flying-fish, it does not seem probable that fishes capable of true flight would have been developed under many subordinate forms, for taking prey of many kinds in many ways, on the land and in the water, until their organs of flight had come to a high stage of perfection, so as to have given them a decided advantage over other animals in the battle for life. Hence the chance of discovering species with transitional grades of structure in a fossil condition will always be less, from their having existed in lesser numbers, than in the case of species with fully developed structures.

I will now give two or three instances both of diversified and of changed habits in the individuals of the same species. In either case it would be easy for natural selection to adapt the structure of the animal to its changed habits, or exclusively to one of its several habits. It is, however, difficult to decide, and immaterial for us, whether habits generally change first and structure afterwards; or whether slight modifications of structure lead to changed habits; both probably often occurring almost simultaneously. Of cases of changed habits it will suffice merely to allude to that of the many British insects which now feed on exotic plants, or exclusively on artificial substances. Of diversified habits innumerable instances could be given: I have often watched a tyrant flycatcher (Saurophagus sulphuratus) in South America, hovering over one spot and then proceeding to another, like a kestrel, and at other times standing stationary on the margin of water, and then dashing into it like a king-fisher at a fish. In our own country the larger titmouse (Parus major) may be seen climbing branches, almost like a creeper; it sometimes, like a shrike, kills small birds by blows on the head; and I have many times seen and heard it hammering the seeds of the yew on a branch, and thus breaking them like a nuthatch. In North America the black bear was seen by Hearne swimming for hours with widely open mouth, thus catching, almost like a whale, insects in the water.

As we sometimes see individuals following habits different from those proper to their species and to the other species of the same genus, we might expect that such individuals would occasionally give rise to new species, having anomalous habits, and with their structure either slightly or considerably modified from that of their type. And such instances occur in nature. Can a more striking instance of adaptation be given than that of a woodpecker for climbing trees and seizing insects in the chinks of the bark? Yet in North America there are woodpeckers which feed largely on fruit, and others with elongated wings which chase insects on the wing. On the plains of La Plata, where hardly a tree grows, there is a woodpecker (Colaptes campestris) which has two toes before and two behind, a long pointed tongue, pointed tail-feathers, sufficiently stiff to support the bird in a vertical position on a post, but not so stiff as in the typical woodpeckers, and a straight strong beak. The beak, however, is not so straight or so strong as in the typical woodpeckers, but it is strong enough to bore into wood. Hence this Colaptes in all the essential parts of its structure is a woodpecker. Even in such trifling characters as the colouring, the harsh tone of the voice, and undulatory flight, its close blood-relationship to our common woodpecker is plainly declared; yet, as I can assert, not only from my own observation, but from those of the accurate Azara, in certain large districts it does not climb trees, and it makes its nest in holes in banks! In certain other districts, however, this same woodpecker, as Mr. Hudson states, frequents trees, and bores holes in the trunk for its nest. I may mention as another illustration of the varied habits of this genus, that a Mexican Colaptes has been described by De Saussure as boring holes into hard wood in order to lay up a store of acorns.

Petrels are the most aërial and oceanic of birds, but in the quiet sounds of Tierra del Fuego, the Puffinuria berardi, in its general habits, in its astonishing power of diving, in its manner of swimming and of flying when made to take flight, would be mistaken by any one for an auk or a grebe; nevertheless it is essentially a petrel, but with many parts of its organisation profoundly modified in relation to its new habits of life; whereas the woodpecker of La Plata has had its structure only slightly modified. In the case of the water-ouzel, the acutest observer by examining its dead body would never have suspected its sub-aquatic habits; yet this bird, which is allied to the thrush family, subsists by diving—using its wings under water, and grasping stones with its feet. All the members of the great order of Hymenopterous insects are terrestrial, excepting the genus Proctotrupes, which Sir John Lubbock has discovered to be aquatic in its habits; it often enters the water and dives about by the use not of its legs but of its wings, and remains as long as four hours beneath the surface; yet it exhibits no modification in structure in accordance with its abnormal habits.

He who believes that each being has been created as we now see it, must occasionally have felt surprise when he has met with an animal having habits and structure not in agreement. What can be plainer than that the webbed feet of ducks and geese are formed for swimming? Yet there are upland geese with webbed feet which rarely go near the water; and no one except Audubon has seen the frigate-bird, which has all its four toes webbed, alight on the surface of the ocean. On the other hand, grebes and coots are eminently aquatic, although their toes are only bordered by membrane. What seems plainer than that the long toes, not furnished with membrane of the Grallatores are formed for walking over swamps and floating plants?—the waterhen and landrail are members of this order, yet the first is nearly as aquatic as the coot, and the second nearly as terrestrial as the quail or partridge. In such cases, and many others could be given, habits have changed without a corresponding change of structure. The webbed feet of the upland goose may be said to have become almost rudimentary in function, though not in structure. In the frigate-bird, the deeply scooped membrane between the toes shows that structure has begun to change.

He who believes in separate and innumerable acts of creation may say, that in these cases it has pleased the Creator to cause a being of one type to take the place of one belonging to another type; but this seems to me only re-stating the fact in dignified language. He who believes in the struggle for existence and in the principle of natural selection, will acknowledge that every organic being is constantly endeavouring to increase in numbers; and that if any one being varies ever so little, either in habits or structure, and thus gains an advantage over some other inhabitant of the same country, it will seize on the place of that inhabitant, however different that may be from its own place. Hence it will cause him no surprise that there should be geese and frigate-birds with webbed feet, living on the dry land and rarely alighting on the water, that there should be long-toed corncrakes, living in meadows instead of in swamps; that there should be woodpeckers where hardly a tree grows; that there should be diving thrushes and diving Hymenoptera, and petrels with the habits of auks.

ORGANS OF EXTREME PERFECTION AND COMPLICATION

To suppose that the eye with all its inimitable contrivances for adjusting the focus to different distances, for admitting different amounts of light, and for the correction of spherical and chromatic aberration, could have been formed by natural selection, seems, I freely confess, absurd in the highest degree. When it was first said that the sun stood still and the world turned round, the common sense of mankind declared the doctrine false; but the old saying of *Vox populi, vox Dei*, as every philosopher knows, cannot be trusted in science. Reason tells me, that if numerous gradations from a simple and imperfect eye to one complex and perfect can be shown to exist, each grade being useful to its possessor, as is certainly the case; if further, the eye ever varies and the variations be inherited, as is likewise certainly the case; and if such variations should be useful to any animal under changing conditions of life, then the difficulty of believing that a perfect and complex eye could be formed by natural selection, though insuperable by our imagination, should not be considered as subversive of the theory. How a nerve comes to be sensitive to light, hardly concerns us more than how life itself originated; but I may remark that, as some of the lowest organisms, in which nerves cannot be detected, are capable of perceiving light, it does not seem impossible that certain elements in their sarcode should become aggregated and developed into nerves, endowed with this special sensibility.

In searching for the gradations through which an organ in any species has been perfected, we ought to look exclusively to its lineal progenitors; but this is scarcely ever possible, and we are forced to look to other species and genera of the same group, that is to the collateral descendants from the same parent-form, in order to see what gradations are possible, and for the chance of some gradations having been transmitted in an unaltered or little altered condition. But the state of the same organ in distinct classes may incidentally throw light on the steps by which it has been perfected.

The simplest organ which can be called an eye consists of an optic nerve, surrounded by pigment-cells and covered by translucent skin, but without any lens or other refractive body. We may, however, according to M. Jourdain, descend even a step lower and find aggregates of pigment-cells, apparently serving as organs of vision, without any nerves, and resting merely on sarcodic tissue. Eyes of the above simple nature are not capable of distinct vision, and serve only to distinguish light from darkness. In certain starfishes, small depressions in the layer of pigment which surrounds the nerve are filled, as described by the author just quoted, with transparent gelatinous matter, projecting with a convex surface, like the cornea in the higher animals. He suggests that this serves not to form an image, but only to concentrate the luminous rays and render their perception more easy. In this concentration of the rays we gain the first and by far the most important step towards the formation of a true, picture-forming eye; for we have only to place the naked extremity of the optic nerve, which in some of the lower animals lies deeply buried in the body, and in some near the surface, at the right distance from the concentrating apparatus, and an image will be formed on it.

In the great class of the Articulata, we may start from an optic nerve simply coated with pigment, the latter sometimes forming a sort of pupil, but destitute of a lens or other optical contrivance. With insects it is now known that the numerous facets on the cornea of their great compound eyes form true lenses, and that the cones include curiously modified nervous filaments. But these organs in the Articulata are so much diversified that Müller formerly made three main classes with seven subdivisions, besides a fourth main class of aggregated simple eyes.

When we reflect on these facts, here given much too briefly, with respect to the wide, diversified, and graduated range of structure in the eyes of the lower animals; and when we bear in mind how small the number of all living forms must be in comparison with those which have become extinct, the difficulty ceases to be very great in believing that natural selection may have converted the simple apparatus of an optic nerve, coated with pigment and invested by transparent membrane, into an optical instrument as perfect as is possessed by any member of the Articulate Class.

He who will go thus far, ought not to hesitate to go one step further, if he finds on finishing this volume that large bodies of facts, otherwise inexplicable, can be explained by the theory of modification through natural selection; he ought to admit that with a structure even as perfect as an eagle's eye might thus be formed, although in this case he does not know the transitional states. It has been objected that in order to modify the eye and still preserve it as a perfect instrument, many changes would have to be effected simultaneously, which, it is assumed, could not be done through natural selection; but as I have attempted to show in my work on the variation of domestic animals, it is not necessary to suppose that the modifications were all simultaneous, if they were extremely slight and gradual. Different kinds of modification would, also, serve for the same general purpose: as Mr. Wallace has remarked, "if a lens has too short or too long a focus, it may be amended either by an alteration of curvature, or an alteration of density; if the curvature be irregular, and the rays do not converge to a point, then any increased regularity of curvature will be an improvement. So the contraction of the iris and the muscular movements of the eye are neither of them essential to vision, but only improvements which might have been added and perfected at any stage of the construction of the instrument." Within the highest division of the animal kingdom, namely, the Vertebrata, we can start from an eye so simple, that it consists, as in the lancelet, of a little sack of transparent skin, furnished with a nerve and lined with pigment, but destitute of any other apparatus. In fishes and reptiles, as Owen has remarked, "the range of gradations of dioptric structures is very great." It is a significant fact that even in man, according to the high authority of Virchow, the beautiful crystalline lens is formed in the embryo by an accumulation of epidermic cells, lying in a sack-like fold of the skin; and the vitreous body is formed from embryonic subcutaneous tissue. To arrive, however, at a just conclusion regarding the formation of the eye, with all its marvellous yet not absolutely perfect characters, it is indispensable that the reason should conquer the imagination; but I have felt the difficulty far too keenly to be surprised at others hesitating to extend the principle of natural selection to so startling a length.

It is scarcely possible to avoid comparing the eye with a telescope. We know that this instrument has been perfected by the long-continued efforts of the highest human intellects; and we naturally infer that the eye has been formed by a somewhat analogous process. But may not this inference be presumptuous? Have we any right to assume that the Creator works by intellectual powers like those of man? If we must compare the eye to an optical instrument, we ought in imagination to take a thick layer of transparent tissue, with spaces filled with fluid, and with a nerve sensitive to light beneath, and then suppose every part of this layer to be continually changing slowly in density, so as to separate into layers of different densities and thicknesses, placed at different distances from each other, and with the surfaces of each layer slowly changing in form. Further we must suppose that there is a power, represented by natural selection or the survival of the fittest, always intently watching each slight alteration in the transparent layers; and carefully preserving each which, under varied circumstances, in any way or in any degree, tends to produce a distincter image. We must suppose each new state of the instrument to be multiplied by the million; each to be preserved until a better one is produced, and then the old ones to be all destroyed. In living bodies, variation will cause the slight alterations, generation will multiply them almost infinitely, and natural selection will pick out with unerring skill each improvement. Let this process go on for millions of years; and during each year on millions of individuals of many kinds; and may we not believe that a living optical instrument might thus be formed as superior to one of glass, as the works of the Creator are to those of man? * * *

SUMMARY: THE LAW OF UNITY OF TYPE AND OF THE CONDITIONS OF
EXISTENCE EMBRACED BY THE THEORY OF NATURAL SELECTION

We have in this chapter discussed some of the difficulties and objections which may be urged against the theory. Many of them are serious; but I think that in the discussion light has been thrown on several facts, which on the belief of independent acts of creation are utterly obscure. We have seen that species at any one period are not indefinitely variable, and are not linked together by a multitude of intermediate gradations, partly because the process of natural selection is always very slow, and at any one time acts only on a few forms; and partly because the very process of natural selection implies the continual supplanting and extinction of preceding and intermediate gradations. Closely allied species, now living on a continuous area, must often have been formed when the area was not continuous, and when the conditions of life did not insensibly graduate away from one part to another. When two varieties are formed in two districts of a continuous area, an intermediate variety will often be formed, fitted for an intermediate zone; but from reasons assigned, the intermediate variety will usually exist in lesser numbers than the two forms which it connects; consequently the two latter, during the course of further modification, from existing in greater numbers, will have a great advantage over the less numerous intermediate variety, and will thus generally succeed in supplanting and exterminating it.

We have seen in this chapter how cautious we should be in concluding that the most different habits of life could not graduate into each other; that a bat, for instance, could not have been formed by natural selection from an animal which at first only glided through the air.

We have seen that a species under new conditions of life may change its habits; or it may have diversified habits, with some very unlike those of its nearest congeners. Hence we can understand, bearing in mind that each organic being is trying to live wherever it can live, how it has arisen that there are upland geese with webbed feet, ground woodpeckers, diving thrushes, and petrels with the habits of auks.

Although the belief that an organ so perfect as the eye could have been formed by natural selection, is enough to stagger any one; yet in the case of any organ, if we know of a long series of gradations in complexity, each good for its possessor, then, under changing conditions of life, there is no logical impossibility in the acquirement of any conceivable degree of perfection through natural selection. In the cases in which we know of no intermediate or transitional states, we should be extremely cautious in concluding that none can have existed, for the metamorphoses of many organs show what wonderful changes in function are at least possible. For instance, a swimbladder has apparently been converted into an air-breathing lung. The same organ having performed simultaneously very different functions, and then having been in part or in whole specialised for one function; and two distinct organs having performed at the same time the same function, the one having been perfected whilst aided by the other, must often have largely facilitated transitions.

We have seen that in two beings widely remote from each other in the natural scale, organs serving for the same purpose and in external appearance closely similar may have been separately and independently formed; but when such organs are closely examined, essential differences in their structure can almost always be detected; and this naturally follows from the principle of natural selection. On the other hand, the common rule throughout nature is infinite diversity of structure for gaining the same end; and this again naturally follows from the same great principle.

In many cases we are far too ignorant to be enabled to assert that a part or organ is so unimportant for the welfare of a species, that modifications in its structure could not have been slowly accumulated by means of natural selection. In many other cases, modifications are probably the direct result of the laws of variation or of growth, independently of any good having been thus gained. But even such structures have often, as we may feel assured, been subsequently taken advantage of, and still further modified, for the good of species under new conditions of life. We may, also, believe that a part formerly of high importance has frequently been retained (as the tail of an aquatic animal by its terrestrial descendants), though it has become of such small importance that it could not, in its present state, have been acquired by means of natural selection.

Natural selection can produce nothing in one species for the exclusive good or injury of another; though it may well produce parts, organs, and excretions highly useful or even indispensable, or again highly injurious to another species, but in all cases at the same time useful to the possessor. In each well-stocked country natural selection acts through the competition of the inhabitants, and consequently leads to success in the battle for life, only in accordance with the standard of that particular country. Hence the inhabitants of one country, generally the smaller one, often yield to the inhabitants of another and generally the larger country. For in the larger country there will have existed more individuals and more diversified forms, and the competition will have been severer, and thus the standard of perfection will have been rendered higher. Natural selection will not necessarily lead to absolute perfection; nor, as far as we can judge by our limited faculties, can absolute perfection be everywhere predicated.

On the theory of natural selection we can clearly understand the full meaning of that old canon in natural history, "Natura non facit saltum." This canon, if we look to the present inhabitants alone of the world, is not strictly correct; but if we include all those of past times, whether known or unknown, it must on this theory be strictly true.

It is generally acknowledged that all organic beings have been formed on two great laws—Unity of Type, and the Conditions of Existence. By unity of type is meant that fundamental agreement in structure which we see in organic beings of the same class, and which is quite independent of their habits of life. On my theory, unity of type is explained by unity of descent. The expression of conditions is fully embraced by the principle of natural selection. For natural selection acts by either now adapting the varying parts of each being to its organic and inorganic conditions of life; or by having adapted them during past periods of time: the adaptations being aided in many cases by the increased use or disuse of parts, being affected by the direct action of the external conditions of life, and subjected in all cases to the several laws of growth and variation. Hence, in fact, the law of the Conditions of Existence is the higher law; as it includes, through the inheritance of former variations and adaptations, that of Unity of Type.

Chapter X

ON THE IMPERFECTION OF THE GEOLOGICAL RECORD

On the absence of intermediate varieties at the present day—On the nature of extinct intermediate varieties; on their number—On the lapse of time, as inferred from the rate of denudation and of deposition—On the lapse of time as estimated by years—On the poorness of our palæontological collections—On the intermittence of geological formations—On the denudation of granitic areas—On the absence of intermediate varieties in any one formation—On the sudden appearance of groups of species—On their sudden appearance in the lowest known fossiliferous strata—Antiquity of the habitable earth.

In the sixth chapter I enumerated the chief objections which might be justly urged against the views maintained in this volume. Most of them have now been discussed. One, namely the distinctness of specific forms, and their not being blended together by innumerable transitional links, is a very obvious difficulty. I assigned reasons why such links do not commonly occur at the present day under the circumstances apparently most favourable for their presence, namely, on an extensive and continuous area with graduated physical conditions. I endeavoured to show, that the life of each species depends in a more important manner on the presence of other already defined organic forms, than on climate, and, therefore, that the really governing conditions of life do not graduate away quite insensibly like heat or moisture. I endeavoured, also, to show that intermediate varieties, from existing in lesser numbers than the forms which they connect, will generally be beaten out and exterminated during the course of further modification and improvement. The main cause, however, of innumerable intermediate links not now occurring everywhere throughout nature, depends on the very process of natural selection, through which new varieties continually take the places of and supplant their parent-forms. But just in proportion as this process of extermination has acted on an enormous scale, so must the number of intermediate varieties, which have formerly existed, be truly enormous. Why then is not every geological formation and every stratum full of such intermediate links? Geology assuredly does not reveal any such finely-graduated organic chain; and this, perhaps, is the most obvious and serious objection which can be urged against the theory. The explanation lies, as I believe, in the extreme imperfection of the geological record.

In the first place, it should always be borne in mind what sort of intermediate forms must, on the theory, have formerly existed. I have found it difficult, when looking at any two species, to avoid picturing to myself forms *directly* intermediate between them. But this is a wholly false view; we should always look for forms intermediate between each species and a common but unknown progenitor; and the progenitor will generally have differed in some respects from all its modified descendants. To give a simple illustration: the fantail and pouter pigeons are both descended from the rock-pigeon; if we possessed all the intermediate varieties which have ever existed, we should have an extremely close series between both and the rock-pigeon; but we should have no varieties directly intermediate between the fantail and pouter; none, for instance, combining a tail somewhat expanded with a crop somewhat enlarged, the characteristic features of these two breeds. These two breeds, moreover, have become so much modified, that, if we had no historical or indirect evidence regarding their origin, it would not have been possible to have determined, from a mere comparison of their structure with that of the rock-pigeon, C. livia, whether they had descended from this species or from some allied form, such as C. œnas.

So, with natural species, if we look to forms very distinct, for instance to the horse and tapir, we have no reason to suppose that links directly intermediate between them ever existed, but between each and an unknown common parent. The common parent will have had in its whole organisation much general resemblance to the tapir and to the horse; but in some points of structure may have differed considerably from both, even perhaps more than they differ from each other. Hence, in all such cases, we should be unable to recognise the parent-form of any two or more species, even if we closely compared the structure of the parent with that of its modified descendants, unless at the same time we had a nearly perfect chain of the intermediate links.

It is just possible by the theory, that one of two living forms might have descended from the other; for instance, a horse from a tapir; and in this case *direct* intermediate links will have existed between them. But such a case would imply that one form had remained for a very long period unaltered, whilst its descendants had undergone a vast amount of change; and the principle of competition between organism and organism, between child and parent, will render this a very rare event; for in all cases the new and improved forms of life tend to supplant the old and unimproved forms.

By the theory of natural selection all living species have been connected with the parent-species of each genus, by differences not greater than we see between the natural and domestic varieties of the same species at the present day; and these parent-species, now generally extinct, have in their turn been similarly connected with more ancient forms; and so on backwards, always converging to the common anestor of each great class. So that the number of intermediate and transitional links, between all living and extinct species, must have been inconceivably great. But assuredly, if this theory be true, such have lived upon the earth.

ON THE LAPSE OF TIME, AS INFERRED FROM THE RATE OF DEPOSITION
AND EXTENT OF DENUDATION

Independently of our not finding fossil remains of such infinitely numerous connecting links, it may be objected that time cannot have sufficed for so great an amount of organic change, all changes having been effected slowly. It is hardly possible for me to recall to the reader who is not a practical geologist, the facts leading the mind feebly to comprehend the lapse of time. He who can read Sir Charles Lyell's grand work on the Principles of Geology, which the future historian will recognise as having produced a revolution in natural science, and yet does not admit how vast have been the past periods of time, may at once close this volume. Not that it suffices to study the Principles of Geology, or to read special treatises by different observers on separate formations, and to mark how each author attempts to give an inadequate idea of the duration of each formation, or even of each stratum. We can best gain some idea of past time by knowing the agencies at work, and learning how deeply the surface of the land has been denuded, and how much sediment has been deposited. As Lyell has well remarked, the extent and thickness of our sedimentary formations are the result and the measure of the denudation which the earth's crust has elsewhere undergone. Therefore a man should examine for himself the great piles of superimposed strata, and watch the rivulets bringing down mud, and the waves wearing away the sea-cliffs, in order to comprehend something about the duration of past time, the monuments of which we see all around us. * * *

ON THE POORNESS OF PALAEONTOLOGICAL COLLECTIONS

Now let us turn to our richest geological museums, and what a paltry display we behold! That our collections are imperfect is admitted by every one. The remark of that admirable palæontologist, Edward Forbes, should never be forgotten, namely, that very many fossil species are known and named from single and often broken specimens, or from a few specimens collected on some one spot. Only a small portion of the surface of the earth has been geologically explored, and no part with sufficient care, as the important discoveries made every year in Europe prove. No organism wholly soft can be preserved. Shells and bones decay and disappear when left on the bottom of the sea, where sediment is not accumulating. We probably take a quite erroneous view, when we assume that sediment is being deposited over nearly the whole bed of the sea, at a rate sufficiently quick to embed and preserve fossil remains. Throughout an enormously large proportion of the ocean, the bright blue tint of the water bespeaks its purity. The many cases on record of a formation conformably covered, after an immense interval of time, by another and later formation, without the underlying bed having suffered in the interval any wear and tear, seem explicable only on the view of the bottom of the sea not rarely lying for ages in an unaltered condition. The remains which do become embedded, if in sand or gravel, will, when the beds are upraised generally be dissolved by the percolation of rain-water charged with carbonic acid. Some of the many kinds of animals which live on the beach between high and low water mark seem to be rarely preserved. For instance, the several species of the Chthamalinæ (a subfamily of sessile cirripedes) coat the rocks all over the world in infinite numbers: they are all strictly littoral, with the exception of a single Mediterranean species, which inhabits deep water, and this has been found fossil in Sicily, whereas not one other species has hitherto been found in any tertiary formation: yet it is known that the genus Chthamalus existed during the Chalk period. Lastly, many great deposits requiring a vast length of time for their accumulation, are entirely destitute of organic remains, without our being able to assign any reason: one of the most striking instances is that of the Flysch formation, which consists of shale and sandstone, several thousand, occasionally even six thousand feet in thickness, and extending for at least 300 miles from Vienna to Switzerland; and although this great mass has been most carefully searched, no fossils, except a few vegetable remains, have been found.

With respect to the terrestrial productions which lived during the Secondary and Palæozoic periods, it is superfluous to state that our evidence is fragmentary in an extreme degree. For instance, until recently not a land-shell was known belonging to either of these vast periods, with the exception of one species discovered by Sir C. Lyell and Dr. Dawson in the carboniferous strata of North America; but now land-shells have been found in the lias. In regard to mammiferous remains, a glance at the historical table published in Lyell's Manual will bring home the truth, how accidental and rare is their preservation, far better than pages of detail. Nor is their rarity surprising, when we remember how large a proportion of the bones of tertiary mammals have been discovered either in caves or in lacustrine deposits; and that not a cave or true lacustrine bed is known belonging to the age of our secondary or palæozoic formations.

But the imperfection in the geological record largely results from another and more important cause than any of the foregoing; namely, from the several formations being separated from each other by wide intervals of time. This doctrine has been emphatically admitted by many geologists and palæontologists, who, like E. Forbes, entirely disbelieve in the change of species. When we see the formations tabulated in written works, or when we follow them in nature, it is difficult to avoid believing that they are closely consecutive. But we know, for instance, from Sir R. Murchison's great work on Russia, what wide gaps there are in that country between the superimposed formations; so it is in North America, and in many other parts of the world. The most skilful geologist, if his attention had been confined exclusively to these large territories, would never have suspected that, during the periods which were blank and barren in his own country, great piles of sediment, charged with new and peculiar forms of life, had elsewhere been accumulated. And if, in each separate territory, hardly any idea can be formed of the length of time which has elapsed between the consecutive formations, we may infer that this could nowhere be ascertained. The frequent and great changes in the mineralogical composition of consecutive formations, generally implying great changes in the geography of the surrounding lands, whence the sediment was derived, accord with the belief of vast intervals of time having elapsed between each formation. * * *

ON THE SUDDEN APPEARANCE OF GROUPS OF ALLIED SPECIES IN THE
LOWEST KNOWN FOSSILIFEROUS STRATA

* * * Those who believe that the geological record is in any degree perfect, will undoubtedly at once reject the theory. For my part, following out Lyell's metaphor, I look at the geological record as a history of the world imperfectly kept, and written in a changing dialect; of this history we possess the last volume alone, relating only to two or three countries. Of this volume, only here and there a short chapter has been preserved; and of each page, only here and there a few lines. Each word of the slowly-changing language, more or less different in the successive chapters, may represent the forms of life, which are entombed in our consecutive formations, and which falsely appear to have been abruptly introduced. On this view, the difficulties above discussed are greatly diminished, or even disappear.

Chapter XV

RECAPITULATION AND CONCLUSION

Recapitulation of the objections to the theory of natural selection —Recapitulation of the general and special circumstances in its favour—Causes of the general belief in the immutability of species —How far the theory of natural selection may be extended—Effects of its adoption on the study of natural history—Concluding remarks.

As this whole volume is one long argument, it may be convenient to the reader to have the leading facts and inferences briefly recapitulated.

That many and serious objections may be advanced against the theory of descent with modification through variation and natural selection, I do not deny. I have endeavored to give to them their full force. Nothing at first can appear more difficult to believe than that the more complex organs and instincts have been perfected, not by means superior to, though analogous with, human reason, but by the accumulation of innumerable slight variations, each good for the individual possessor. Nevertheless, this difficulty, though appearing to our imagination insuperably great, cannot be considered real if we admit the following propositions, namely, that all parts of the organisation and instincts offer, at least, individual differences—that there is a struggle for existence leading to the preservation of profitable deviations of structure or instinct—and, lastly, that gradations in the state of perfection of each organ may have existed, each good of its kind. The truth of these propositions cannot, I think, be disputed.

It is, no doubt, extremely difficult even to conjecture by what gradations many structures have been perfected, more especially amongst broken and failing groups of organic beings, which have suffered much extinction, but we see so many strange gradations in nature, that we ought to be extremely cautious in saying that any organ or instinct, or any whole structure, could not have arrived at its present state by many graduated steps. There are, it must be admitted, cases of special difficulty opposed to the theory of natural selection; and one of the most curious of these is the existence in the same community of two or three defined castes of workers or sterile female ants; but I have attempted to show how these difficulties can be mastered.

With respect to the almost universal sterility of species when first crossed, which forms so remarkable a contrast with the almost universal fertility of varieties when crossed, I must refer the reader to the recapitulation of the facts given at the end of the ninth chapter, which seem to me conclusively to show that this sterility is no more a special endowment than is the incapacity of two distinct kinds of trees to be grafted together; but that it is incidental on differences confined to the reproductive systems of the intercrossed species. We see the truth of this conclusion in the vast difference in the results of crossing the same two species reciprocally,—that is, when one species is first used as the father and then as the mother. Analogy from the consideration of dimorphic and trimorphic plants clearly leads to the same conclusion, for when the forms are illegitimately united, they yield few or no seed, and their offspring are more or less sterile; and these forms belong to the same undoubted species, and differ from each other in no respect except in their reproductive organs and functions.

Although the fertility of varieties when intercrossed and of their mongrel offspring has been asserted by so many authors to be universal, this cannot be considered as quite correct after the facts given on the high authority of Gärtner and Kölreuter. Most of the varieties which have been experimented on have been produced under domestication; and as domestication (I do not mean mere confinement) almost certainly tends to eliminate that sterility which, judging from analogy, would have affected the parent-species if intercrossed, we ought not to expect that domestication would likewise induce sterility in their modified descendants when crossed. This elimination of sterility apparently follows from the same cause which allows our domestic animals to breed freely under diversified circumstances; and this again apparently follows from their having been gradually accustomed to frequent changes in their conditions of life.

A double and parallel series of facts seems to throw much light on the sterility of species, when first crossed, and of their hybrid offspring. On the one side, there is good reason to believe that slight changes in the conditions of life give vigour and fertility to all organic beings. We know also that a cross between the distinct individuals of the same variety, and between distinct varieties, increases the number of their offspring, and certainly gives to them increased size and vigour. This is chiefly owing to the forms which are crossed having been exposed to somewhat different conditions of life; for I have ascertained by a laborious series of experiments that if all the individuals of the same variety be subjected during several generations to the same conditions, the good derived from crossing is often much diminished or wholly disappears. This is one side of the case. On the other side, we know that species which have long been exposed to nearly uniform conditions, when they are subjected under confinement to new and greatly changed conditions, either perish, or if they survive, are rendered sterile, though retaining perfect health. This does not occur, or only in a very slight degree, with our domesticated productions, which have long been exposed to fluctuating conditions. Hence when we find that hybrids produced by a cross between two distinct species are few in number, owing to their perishing soon after conception or at a very early age, or if surviving that they are rendered more or less sterile, it seems highly probable that this result is due to their having been in fact subjected to a great change in their conditions of life, from being compounded of two distinct organisations. He who will explain in a definite manner why, for instance, an elephant or a fox will not breed under confinement in its native country, whilst the domestic pig or dog will breed freely under the most diversified conditions, will at the same time be able to give a definite answer to the question why two distinct species, when crossed, as well as their hybrid offspring, are generally rendered more or less sterile, whilst two domesticated varieties when crossed and their mongrel offspring are perfectly fertile.

Turning to geographical distribution, the difficulties encountered on the theory of descent with modification are serious enough. All the individuals of the same species, and all the species of the same genus, or even higher group, are descended from common parents; and therefore, in however distant and isolated parts of the world they may now be found, they must in the course of successive generations have travelled from some one point to all the others. We are often wholly unable even to conjecture how this could have been effected. Yet, as we have reason to believe that some species have retained the same specific form for very long periods of time, immensely long as measured by years, too much stress ought not to be laid on the occasional wide diffusion of the same species; for during very long periods there will always have been a good chance for wide migration by many means. A broken or interrupted range may often be accounted for by the extinction of the species in the intermediate regions. It cannot be denied that we are as yet very ignorant as to the full extent of the various climatal and geographical changes which have affected the earth during modern periods; and such changes will often have facilitated migration. As an example, I have attempted to show how potent has been the influence of the Glacial period on the distribution of the same and of allied species throughout the world. We are as yet profoundly ignorant of the many occasional means of transport. With respect to distinct species of the same genus inhabiting distant and isolated regions, as the process of modification has necessarily been slow, all the means of migration will have been possible during a very long period; and consequently the difficulty of the wide diffusion of the species of the same genus is in some degree lessened.

As according to the theory of natural selection an interminable number of intermediate forms must have existed, linking together all the species in each group by gradations as fine are our existing varieties, it may be asked: Why do we not see these linking forms all around us? Why are not all organic beings blended together in an inextricable chaos? With respect to existing forms, we should remember that we have no right to expect (excepting in rare cases) to discover *directly* connecting links between them, but only between each and some extinct and supplanted form. Even on a wide area, which has during a long period remained continuous, and of which the climatic and other conditions of life change insensibly in proceeding from a district occupied by one species into another district occupied by a closely allied species, we have no just right to expect often to find intermediate varieties in the intermediate zones. For we have reason to believe that only a few species of a genus ever undergo change; the other species becoming utterly extinct and leaving no modified progeny. Of the species which do change, only a few within the same country change at the same time; and all modifications are slowly effected. I have also shown that the intermediate varieties which probably at first existed in the intermediate zones, would be liable to be supplanted by the allied forms on either hand; for the latter, from existing in greater numbers, would generally be modified and improved at a quicker rate than the intermediate varieties, which existed in lesser numbers; so that the intermediate varieties would, in the long run, be supplanted and exterminated.

On this doctrine of the extermination of an infinitude of connecting links, between the living and extinct inhabitants of the world, and at each successive period between the extinct and still older species, why is not every geological formation charged with such links? Why does not every collection of fossil remains afford plain evidence of the gradation and mutation of the forms of life? Although geological research has undoubtedly revealed the former existence of many links, bringing numerous forms of life much closer together, it does not yield the infinitely many fine gradations between past and present species required on the theory; and this is the most obvious of the many objections which may be urged against it. Why, again, do whole groups of allied species appear, though this appearance is often false, to have come in suddenly on the successive geological stages? Although we now know that organic beings appeared on this globe, at a period incalculably remote, long before the lowest bed of the Cambrian system was deposited, why do we not find beneath this system great piles of strata stored with the remains of the progenitors of the Cambrian fossils? For on the theory, such strata must somewhere have been deposited at these ancient and utterly unknown epochs of the world's history.

I can answer these questions and objections only on the supposition that the geological record is far more imperfect than most geologists believe. The number of specimens in all our museums is absolutely as nothing compared with the countless generations of countless species which have certainly existed. The parent-form of any two or more species would not be in all its characters directly intermediate between its modified offspring, any more than the rock-pigeon is directly intermediate in crop and tail between its descendants, the pouter and fantail pigeons. We should not be able to recognise a species as the parent of another and modified species, if we were to examine the two ever so closely, unless we possessed most of the intermediate links; and owing to the imperfection of the geological record, we have no just right to expect to find so many links. If two or three, or even more linking forms were discovered, they would simply be ranked by many naturalists as so many new species, more especially if found in different geological

sub-stages, let their differences be ever so slight. Numerous existing doubtful forms could be named which are probably varieties;
but who will pretend that in future ages so many fossil links will
be discovered, that naturalists will be able to decide whether or not
these doubtful forms ought to be called varieties? Only a small
portion of the world has been geologically explored. Only organic
beings of certain classes can be preserved in a fossil condition, at
least in any great number. Many species when once formed never
undergo any further change but become extinct without leaving
modified descendants; and the periods, during which species have
undergone modification, though long as measured by years, have
probably been short in comparison with the periods during which
they retain the same form. It is the dominant and widely ranging
species which vary most frequently and vary most, and varieties are
often at first local—both causes rendering the discovery of intermediate links in any one formation less likely. Local varieties will
not spread into other and distant regions until they are considerably modified and improved; and when they have spread, and are
discovered in a geological formation, they appear as if suddenly
created there, and will be simply classed as new species. Most
formations have been intermittent in their accumulation; and
their duration has probably been shorter than the average duration
of specific forms. Successive formations are in most cases separated
from each other by blank intervals of time of great length; for
fossiliferous formations thick enough to resist future degradations
can as a general rule be accumulated only where much sediment is
deposited on the subsiding bed of the sea. During the alternate
periods of elevation and of stationary level the record will generally
be blank. During these latter periods there will probably be more
variability in the forms of life; during periods of subsidence, more
extinction.

With respect to the absence of strata rich in fossils beneath the
Cambrian formation, I can recur only to the hypothesis given in
the tenth chapter; namely, that though our continents and oceans
have endured for an enormous period in nearly their present relative positions, we have no reason to assume that this has always
been the case; consequently formations much older than any now
known may lie buried beneath the great oceans. With respect to
the lapse of time not having been sufficient since our planet was
consolidated for the assumed amount of organic change, and this
objection, as urged by Sir William Thompson, is probably one of
the gravest as yet advanced, I can only say, firstly, that we do not
know at what rate species change as measured by years, and secondly, that many philosophers are not as yet willing to admit that
we know enough of the constitution of the universe and of the in-

terior of our globe to speculate with safety on its past duration.

That the geological record is imperfect all will admit; but that it is imperfect to the degree required by our theory, few will be inclined to admit. If we look to long enough intervals of time, geology plainly declares that species have all changed; and they have changed in the manner required by the theory, for they have changed slowly and in a graduated manner. We clearly see this in the fossil remains from consecutive formations invariably being much more closely related to each other, than are the fossils from widely separated formations.

Such is the sum of the several chief objections and difficulties which may be justly urged against the theory; and I have now briefly recapitulated the answers and explanations which, as far as I can see, may be given. I have felt these difficulties far too heavily during many years to doubt their weight. But it deserves especial notice that the more important objections relate to questions on which we are confessedly ignorant; nor do we know how ignorant we are. We do not know all the possible transitional gradations between the simplest and the most perfect organs; it cannot be pretended that we know all the varied means of distribution during the long lapse of years, or that we know how imperfect is the Geological Record. Serious as these several objections are, in my judgment they are by no means sufficient to overthrow the theory of descent with subsequent modification.

Now let us turn to the other side of the argument. Under domestication we see much variability, caused, or at least excited, by changed conditions of life; but often in so obscure a manner, that we are tempted to consider the variations as spontaneous. Variability is governed by many complex laws,—by correlated growth, compensation, the increased use and disuse of parts, and the definite action of the surrounding conditions. There is much difficulty in ascertaining how largely our domestic productions have been modified; but we may safely infer that the amount has been large, and that modifications can be inherited for long periods. As long as the conditions of life remain the same, we have reason to believe that a modification, which has already been inherited for many generations, may continue to be inherited for an almost infinite number of generations. On the other hand, we have evidence that variability when it has once come into play, does not cease under domestication for a very long period; nor do we know that it ever ceases, for new varieties are still occasionally produced by our oldest domesticated productions.

Variability is not actually caused by man; he only unintentionally exposes organic beings to new conditions of life, and then nature acts on the organisation and causes it to vary. But man can and does select the variations given to him by nature, and thus accumulates them in any desired manner. He thus adapts animals and plants for his own benefit or pleasure. He may do this methodically, or he may do it unconsciously by preserving the individuals most useful or pleasing to him without any intention of altering the breed. It is certain that he can largely influence the character of a breed by selecting, in each successive generation, individual differences so slight as to be inappreciable except by an educated eye. This unconscious process of selection has been the great agency in the formation of the most distinct and useful domestic breeds. That many breeds produced by man have to a large extent the character of natural species, is shown by the inextricable doubts whether many of them are varieties or aboriginally distinct species.

There is no reason why the principles which have acted so efficiently under domestication should not have acted under nature. In the survival of favoured individuals and races, during the constantly-recurrent Struggle for Existence, we see a powerful and ever-acting form of Selection. The struggle for existence inevitably follows from the high geometrical ratio of increase which is common to all organic beings. This high rate of increase is proved by calculation,—by the rapid increase of many animals and plants during a succession of peculiar seasons, and when naturalised in new countries. More individuals are born than can possibly survive. A grain in the balance may determine which individuals shall live and which shall die,—which variety or species shall increase in number, and which shall decrease, or finally become extinct. As the individuals of the same species come in all respects into the closest competition with each other, the struggle will generally be most severe between them; it will be almost equally severe between the varieties of the same species, and next in severity between the species of the same genus. On the other hand the struggle will often be severe between beings remote in the scale of nature. The slightest advantage in certain individuals, at any age or during any season, over those with which they come into competition, or better adaptation in however slight a degree to the surrounding physical conditions, will, in the long run, turn the balance.

With animals having separated sexes, there will be in most cases a struggle between the males for the possession of the females. The most vigorous males, or those which have most successfully struggled with their conditions of life, will generally leave most progeny. But success will often depend on the males having special weapons, or means of defense, or charms; and a slight advantage will lead to victory.

As geology plainly proclaims that each land has undergone

great physical changes, we might have expected to find that organic beings have varied under nature, in the same way as they have varied under domestication. And if there has been any variability under nature, it would be an unaccountable fact if natural selection had not come into play. It has often been asserted, but the assertion is incapable of proof, that the amount of variation under nature is a strictly limited quantity. Man, though acting on external characters alone and often capriciously, can produce within a short period a great result by adding up mere individual differences in his domestic productions; and every one admits that species present individual differences. But, besides such differences, all naturalists admit that natural varieties exist, which are considered sufficiently distinct to be worthy of record in systematic works. No one has drawn any clear distinction between individual differences and slight varieties; or between more plainly marked varieties and sub-species, and species. On separate continents, and on different parts of the same continent when divided by barriers of any kind, and on outlying islands, what a multitude of forms exist, which some experienced naturalists rank as varieties, others as geographical races or sub-species, and others as distinct, though closely allied species!

If then, animals and plants do vary, let it be ever so slightly or slowly, why should not variations or individual differences, which are in any way beneficial, be preserved and accumulated through natural selection, or the survival of the fittest? If man can by patience select variations useful to him, why, under changing and complex conditions of life, should not variations useful to nature's living products often arise, and be preserved or selected? What limit can be put to this power, acting during long ages and rigidly scrutinising the whole constitution, structure, and habits of each creature,—favouring the good and rejecting the bad? I can see no limit to this power, in slowly and beautifully adapting each form to the most complex relations of life. The theory of natural selection, even if we look no farther than this, seems to be in the highest degree probable. I have already recapitulated, as fairly as I could, the opposed difficulties and objections: now let us turn to the special facts and arguments in favour of the theory.

On the view that species are only strongly marked and permanent varieties, and that each species first existed as a variety, we can see why it is that no line of demarcation can be drawn between species, commonly supposed to have been produced by special acts of creation, and varities which are acknowledged to have been produced by secondary laws. On this same view we can understand how it is that in a region where many species of a genus have been produced, and where they now flourish, these same species should present many varieties; for where the manufactory of species has been active, we might expect, as a general rule, to find it still in action; and this is the case if varieties be incipient species. Moreover, the species of the larger genera, which afford the greater number of varieties or incipient species, retain to a certain degree the character of varieties; for they differ from each other by a less amount of difference than do the species of smaller genera. The closely allied species also of the larger genera apparently have restricted ranges, and in their affinities they are clustered in little groups round other species—in both respects resembling varieties. These are strange relations on the view that each species was independently created, but are intelligible if each existed first as a variety.

As each species tends by its geometrical rate of reproduction to increase inordinately in number; and as the modified descendants of each species will be enabled to increase by as much as they become more diversified in habits and structure, so as to be able to seize on many and widely different places in the economy of nature, there will be a constant tendency in natural selection to preserve the most divergent offspring of any one species. Hence, during a long-continued course of modification, the slight differences characteristic of varieties of the same species, tend to be augmented into the greater differences characteristic of the species of the same genus. New and improved varieties will inevitably supplant and exterminate the older, less improved, and intermediate varieties; and thus species are rendered to a large extent defined and distinct objects. Dominant species belonging to the larger groups within each class tend to give birth to new and dominant forms; so that each large group tends to become still larger, and at the same time more divergent in character. But as all groups cannot thus go on increasing in size, for the world would not hold them, the more dominant groups beat the less dominant. This tendency in the large groups to go on increasing in size and diverging in character, together with the inevitable contingency of much extinction, explains the arrangement of all the forms of life in groups subordinate to groups, all within a few great classes, which has prevailed throughout all time. This grand fact of the grouping of all organic beings under what is called the Natural System, is utterly inexplicable on the theory of creation.

As natural selection acts solely by accumulating slight, successive, favourable variations, it can produce no great or sudden modifications; it can act only by short and slow steps. Hence, the canon of "Natura non facit saltum," which every fresh addition to our knowledge tends to confirm, is on this theory intelligible. We can see why throughout nature the same general end is gained by an almost infinite diversity of means, for every peculiarity when once acquired is long inherited, and structures already modified in many different ways have to be adapted for the same general purpose. We can, in short, see why nature is prodigal in variety, though niggard in innovation. But why this should be a law of nature if each species has been independently created no man can explain.

Many other facts are, as it seems to me, explicable on this theory. How strange it is that a bird, under the form of a woodpecker, should prey on insects on the ground; that upland geese which rarely or never swim, should possess webbed feet; that a thrush-like bird should dive and feed on sub-aquatic insects; and that a petrel should have the habits and structure fitting it for the life of an auk! and so in endless other cases. But on the view of each species constantly trying to increase in number, with natural selection always ready to adapt the slowly varying descendants of each to any unoccupied or ill-occupied place in nature, these facts cease to be strange, or might even have been anticipated.

We can to a certain extent understand how it is that there is so much beauty throughout nature; for this may be largely attributed to the agency of selection. That beauty, according to our sense of it, is not universal, must be admitted by every one who will look at some venomous snakes, at some fishes, and at certain hideous bats with a distorted resemblance to the human face. Sexual selection has given the most brilliant colours, elegant patterns, and other ornaments to the males, and sometimes to both sexes of many birds, butterflies, and other animals. With birds it has often rendered the voice of the male musical to the female, as well as to our ears. Flowers and fruit have been rendered conspicuous by brilliant colours in contrast with the green foliage, in order that the flowers may be readily seen, visited and fertilised by insects, and the seeds disseminated by birds. How it comes that certain colours, sounds, and forms should give pleasure to man and the lower animals,—that is, how the sense of beauty in its simplest form was first acquired,—we do not know any more than how certain odours and flavours were first rendered agreeable.

As natural selection acts by competition, it adapts and improves the inhabitants of each country only in relation to their co-inhabitants; so that we need feel no surprise at the species of any one country, although on the ordinary view supposed to have been created and specially adapted for that country, being beaten and supplanted by the naturalised productions from another land. Nor ought we to marvel if all the contrivances in nature be not, as far as we can judge, absolutely perfect, as in the case even of the human eye; or if some of them be abhorrent to our ideas of fitness. We need not marvel at the sting of the bee, when used against an enemy, causing the bee's own death; at drones being produced in such great numbers for one single act, and being then slaughtered by their sterile sisters; at the astonishing waste of pollen by our fir-trees; at the instinctive hatred of the queenbee for her own fertile daughters; at the ichneumonidæ feeding within the living bodies of caterpillars; or at other such cases. The wonder indeed is, on the theory of natural selection, that more cases of the want of absolute perfection have not been detected.

The complex and little known laws governing the production of varieties are the same, as far as we can judge, with the laws which have governed the production of distinct species. In both cases physical conditions seem to have produced some direct and definite effect, but how much we cannot say. Thus, when varieties enter any new station, they occasionally assume some of the characters proper to the species of that station. With both varieties and species, use and disuse seem to have produced a considerable effect; for it is impossible to resist this conclusion when we look, for instance, at the logger-headed duck, which has wings incapable of flight, in nearly the same condition as in the domestic duck; or when we look at the burrowing tucu-tucu, which is occasionally blind, and then at certain moles, which are habitually blind and have their eyes covered with skin; or when we look at the blind animals inhabiting the dark caves of America and Europe. With varieties and species, correlated variation seems to have played an important part, so that when one part has been modified other parts have been necessarily modified. With both varieties and species, reversions to long-lost characters occasionally occur. How inexplicable on the theory of creation is the occasional appearance of stripes on the shoulders and legs of the several species of the horse-genus and of their hybrids! How simply is this fact explained if we believe that these species are all descended from a striped progenitor, in the same manner as the several domestic breeds of the pigeon are descended from the blue and barred rock-pigeon!

On the ordinary view of each species having been independently created, why should specific characters, or those by which the species of the same genus differ from each other, be more variable than generic characters in which they all agree? Why, for instance, should the colour of a flower be more likely to vary in any one species of a genus, if the other species possess differently coloured flowers, than if all possessed the same coloured flowers? If species are only well-marked varieties, of which the characters have become in a high degree permanent, we can understand this fact; for they have already varied since they branched off from a common progenitor in certain characters, by which they have come to be specifically distinct from each other; therefore these same characters would be more likely again to vary than the generic characters which have been inherited without change for an immense period. It is inexplicable on the theory of creation why a part developed in a very unusual manner in one species alone of a genus, and therefore, as we may naturally infer, of great importance to that species, should be eminently liable to variation; but, on our view, this part has undergone, since the several species branched off from a common progenitor, an unusual amount of variability and modification, and therefore we might expect the part generally to be still variable. But a part may be developed in the most unusual manner, like the wing of a bat, and yet not be more variable than any other structure, if the part be common to many subordinate forms, that is, if it has been inherited for a very long period; for in this case, it will have been rendered constant by long-continued natural selection.

Glancing at instincts, marvellous as some are, they offer no greater difficulty than do corporeal structures on the theory of the natural selection of successive slight, but profitable modifications. We can thus understand why nature moves by graduated steps in endowing different animals of the same class with their several instincts. I have attempted to show how much light the principle of gradation throws on the admirable architectural powers of the hive-bee. Habit no doubt often comes into play in modifying instincts; but it certainly is not indispensable, as we see in the case of neuter insects, which leave no progeny to inherit the effects of long-continued habit. On the view of all the species of the same genus having descended from a common parent, and having inherited much in common, we can understand how it is that allied species, when placed under widely different conditions of life, yet follow nearly the same instincts; why the thrushes of tropical and temperate South America, for instance, line their nests with mud like our British species. On the view of instincts having been slowly acquired through natural selection, we need not marvel at some instincts being not perfect and liable to mistakes, and at many instincts causing other animals to suffer.

If species be only well-marked and permanent varieties, we can at once see why their crossed offspring should follow the same complex laws in their degrees and kinds of resemblance to their parents,—in being absorbed into each other by successive crosses, and in other such points,—as do the crossed offspring of acknowledged varieties. This similarity would be a strange fact, if species had been independently created and varieties had been produced through secondary laws.

If we admit that the geological record is imperfect to an extreme degree, then the facts, which the record does give, strongly support the theory of descent with modifications. New species have come on the stage slowly and at successive intervals; and the amount of change, after equal intervals of time, is widely different in different groups. The extinction of species and of whole groups of species which has played so conspicuous a part in the history of the organic world, almost inevitably follows from the principle of natural selection; for old forms are supplanted by new and improved forms. Neither single species nor groups of species reappear when the chain of ordinary generation is once broken. The gradual diffusion of dominant forms, with the slow modification of their descendants, causes the forms of life, after long intervals of time, to appear as if they had changed simultaneously throughout the world. The fact of the fossil remains of each formation being in some degree intermediate in character between the fossils in the formations above and below, is simply explained by their intermediate position in the chain of descent. The grand fact that all extinct beings can be classed with all recent beings, naturally follows from the living and the extinct being the offspring of common parents. As species have generally diverged in character during their long course of descent and modification, we can understand why it is that the more ancient forms, or early progenitors of each group, so often occupy a position in some degree intermediate between existing groups. Recent forms are generally looked upon as being, on the whole, higher in the scale of organisation than ancient forms; and they must be higher, in so far as the later and more improved forms have conquered the older and less improved forms in the struggle for life; they have also generally had their organs more specialised for different functions. This fact is perfectly compatible with numerous beings still retaining simple and but little improved structures, fitted for simple conditions of life; it is likewise compatible with some forms having retrograded in organisation, by having become at each stage of descent better fitted for new and degraded habits of life. Lastly, the wonderful law of the long endurance of allied forms on the same continent,—of marsupials in Australia, of edentata in America, and other such cases,—is intelligible, for within the same country the existing and the extinct will be closely allied by descent.

Looking to geographical distribution, if we admit that there has been during the long course of ages much migration from one part of the world to another, owing to former climatal and geographical changes and to the many occasional and unknown means of dispersal, then we can understand, on the theory of descent with modification, most of the great leading facts in Distribution. We can see why there should be so striking a parallelism in the distribution of organic beings throughout space, and in their geological succession throughout time; for in both cases the beings have been connected by the bond of ordinary generation, and the means of modification have been the same. We see the full meaning of the wonderful fact, which has struck every traveller, namely, that on the same continent, under the most diverse conditions, under heat and cold, on mountain and lowland, on deserts and marshes, most of the inhabitants within each great class are plainly related; for they are the descendants of the same progenitors and early colonists. On this same principle of former migration, combined in most cases with modification, we can understand, by the aid of the Glacial period, the identity of some few plants, and the close alliance of many others, on the most distant mountains, and in the northern and southern temperate zones; and likewise the close alliance of some of the inhabitants of the sea in the northern and southern temperate latitudes, though separated by the whole intertropical ocean. Although two countries may present physical conditions as closely similar as the same species ever require, we need feel no surprise at their inhabitants being widely different, if they have been for a long period completely sundered from each other; for as the relation of organism to organism is the most important of all relations, and as the two countries will have received colonists at various periods and in different proportions, from some other country or from each other, the course of modification in the two areas will inevitably have been different.

On this view of migration, with subsequent modification, we see why oceanic islands are inhabited by only few species, but of these, why many are peculiar or endemic forms. We clearly see why species belonging to those groups of animals which cannot cross wide spaces of the ocean, as frogs and terrestrial mammals, do not inhabit oceanic islands; and why, on the other hand, new and peculiar species of bats, animals which can traverse the ocean, are found on islands far distant from any continent. Such cases as the presence of peculiar species of bats on oceanic islands and the absence of all other terrestrial mammals, are facts utterly inexplicable on the theory of independent acts of creation.

The existence of closely allied or representative species in any two areas, implies, on the theory of descent with modification, that the same parent-forms formerly inhabited both areas; and we almost invariably find that wherever many closely allied species inhabit two areas, some identical species are still common to both. Wherever many closely allied yet distinct species occur, doubtful forms and varieties belonging to the same groups likewise occur. It is a rule of high generality that the inhabitants of each area are related to the inhabitants of the nearest source whence immigrants might have been derived. We see this in the striking relation of nearly all plants and animals of the Galapagos archipelago, of Juan

Fernandez, and of the other American islands, to the plants and animals of the neighbouring American mainland; and of those of the Cape de Verde Archipelago, and of the other African islands to the African mainland. It must be admitted that these facts receive no explanation on the theory of creation.

The fact, as we have seen, that all past and present organic beings can be arranged within a few great classes, in groups subordinate to groups, and with the extinct groups often falling in between the recent groups, is intelligible on the theory of natural selection with its contingencies of extinction and divergence of character. On these same principles we see how it is, that the mutual affinities of the forms within each class are so complex and circuitous. We see why certain characters are far more serviceable than others for classification;—why adaptive characters, though of paramount importance to the beings, are of hardly any importance in classification; why characters derived from rudimentary parts, though of no service to the beings, are often of high classificatory value; and why embryological characters are often the most valuable of all. The real affinities of all organic beings, in contradistinction to their adaptive resemblances, are due to inheritance or community of descent. The Natural System is a genealogical arrangement, with the acquired grades of difference, marked by the terms, varieties, species, genera, families, &c.; and we have to discover the lines of descent by the most permanent characters whatever they may be and of however slight vital importance.

The similar framework of bones in the hand of a man, wing of a bat, fin of the porpoise, and leg of the horse,—the same number of vertebræ forming the neck of the giraffe and of the elephant,—and innumerable other such facts, at once explain themselves on the theory of descent with slow and slight successive modifications. The similarity of pattern in the wing and in the leg of a bat, though used for such different purpose,—in the jaws and legs of a crab,—in the petals, stamens, and pistils of a flower, is likewise, to a large extent, intelligible on the view of the gradual modification of parts or organs, which were aboriginally alike in an early progenitor in each of these classes. On the principle of successive variations not always supervening at an early age, and being inherited at a corresponding not early period of life, we clearly see why the embryos of mammals, birds, reptiles, and fishes should be so closely similar, and so unlike the adult forms. We may cease marvelling at the embryo of an airbreathing mammal or bird having branchial slits and arteries running in loops, like those of a fish which has to breathe the air dissolved in water by the aid of well-developed branchiæ.

Disuse, aided sometimes by natural selection, will often have reduced organs when rendered useless under changed habits or conditions of life; and we can understand on this view the meaning of rudimentary organs. But disuse and selection will generally act on each creature, when it has come to maturity and has to play its full part in the struggle for existence, and will thus have little power on an organ during early life; hence the organ will not be reduced or rendered rudimentary at this early age. The calf, for instance, has inherited teeth, which never cut through the gums of the upper jaw, from an early progenitor having well-developed teeth; and we may believe, that the teeth in the mature animal were formerly reduced by disuse, owing to the tongue and palate, or lips, having become excellently fitted through natural selection to browse without their aid; whereas in the calf, the teeth have been left unaffected, and on the principle of inheritance at corresponding ages have been inherited from a remote period to the present day. On the view of each organism with all its separate parts having been specially created, how utterly inexplicable is it that organs bearing the plain stamp of inutility, such as the teeth in the embryonic calf or the shrivelled wings under the soldered wing-covers of many beetles, should so frequently occur. Nature may be said to have taken pains to reveal her scheme of modification, by means of rudimentary organs, of embryological and homologous structures, but we are too blind to understand her meaning.

I have now recapitulated the facts and considerations which have thoroughly convinced me that species have been modified, during a long course of descent. This has been effected chiefly through the natural selection of numerous successive, slight, favourable variations; aided in an important manner, by the inherited effects of the use and disuse of parts; and in an unimportant manner, that is in relation to adaptive structures, whether past or present, by the direct action of external conditions, and by variations which seem to us in our ignorance to arise spontaneously. It appears that I formerly underrated the frequency and value of these latter forms of variation, as leading to permanent modifications of structure independently of natural selection. But as my conclusions have lately been much misrepresented, and it has been stated that I attribute the modification of species exclusively to natural selection, I may be permitted to remark that in the first edition of this work, and subsequently, I placed in a most conspicuous position—namely, at the close of the Introduction—the following words: "I am convinced that natural selection has been the main but not the exclusive means of modification." This has been of no avail. Great is the power of steady misrepresentation; but the history of science shows that fortunately this power does not long endure.

It can hardly be supposed that a false theory would explain, in so satisfactory a manner as does the theory of natural selection, the several large classes of facts above specified. It has recently been objected that this is an unsafe method of arguing; but it is a method used in judging of the common events of life, and has often been used by the greatest natural philosophers. The undulatory theory of light has thus been arrived at; and the belief in the revolution of the earth on its own axis was until lately supported by hardly any direct evidence. It is no valid objection that science as yet throws no light on the far higher problem of the essence or origin of life. Who can explain what is the essence of the attraction of gravity? No one now objects to following out the results consequent on this unknown element of attraction; notwithstanding that Leibnitz formerly accused Newton of introducing "occult qualities and miracles into philosophy."

I see no good reason why the views given in this volume should shock the religious feelings of any one. It is satisfactory, as showing how transient such impressions are, to remember that the greatest discovery ever made by man, namely, the law of the attraction of gravity, was also attacked by Leibnitz, "as subversive of natural, and inferentially of revealed, religion." A celebrated author and divine has written to me that "he has gradually learnt to see that it is just as noble a conception of the Deity to believe that He created a few original forms capable of self-development into other and needful forms, as to believe that He required a fresh act of creation to supply the voids caused by the action of His laws."

Why, it may be asked, until recently did nearly all the most eminent living naturalists and geologists disbelieve in the mutability of species? It cannot be asserted that organic beings in a state of nature are subject to no variation; it cannot be proved that the amount of variation in the course of long ages is a limited quality; no clear distinction has been, or can be, drawn between species and well-marked varieties. It cannot be maintained that species when intercrossed are invariably sterile, and varieties invariably fertile; or that sterility is a special endowment and sign of creation. The belief that species were immutable productions was almost unavoidable as long as the history of the world was thought to be of short duration; and now that we have acquired some idea of the lapse of time, we are too apt to assume, without proof, that the geological record is so perfect that it would have afforded us plain evidence of the mutation of species, if they had undergone mutation.

But the chief cause of our natural unwillingness to admit that one species has given birth to clear and distinct species, is that we are always slow in admitting great changes of which we do not see the steps. The difficulty is the same as that felt by so many geologists, when Lyell first insisted that long lines of inland cliffs had been formed, the great valleys excavated, by the agencies which we see still at work. The mind cannot possibly grasp the full meaning of the term of even a million years; it cannot add up and perceive the full effects of many slight variations, accumulated during an almost infinite number of generations.

Although I am fully convinced of the truth of the views given in this volume under the form of an abstract, I by no means expect to convince experienced naturalists whose minds are stocked with a multitude of facts all viewed, during a long course of years, from a point of view directly opposite to mine. It is so easy to hide our ignorance under such expressions as the "plan of creation," "unity of design," &c., and to think that we give an explanation when we only re-state a fact. Any one whose disposition leads him to attach more weight to unexplained difficulties than to the explanation of a certain number of facts will certainly reject the theory. A few naturalists, endowed with much flexibility of mind, and who have already begun to doubt the immutability of species, may be influenced by this volume; but I look with confidence to the future,— to young and rising naturalists, who will be able to view both sides of the question with impartiality. Whoever is led to believe that species are mutable will do good service by conscientiously expressing his conviction; for thus only can the load of prejudice by which this subject is overwhelmed be removed.

Several eminent naturalists have of late published their belief that a multitude of reputed species in each genus are not real species; but that other species are real, that is, have been independently created. This seems to me a strange conclusion to arrive at. They admit that a multitude of forms, which till lately they themselves thought were special creations, and which are still thus looked at by the majority of naturalists, and which consequently have all the external characteristic features of true species,—they admit that these have been produced by variation, but they refuse to extend the same view to other and slightly different forms. Nevertheless they do not pretend that they can define, or even conjecture, which are the created forms of life, and which are those produced by secondary laws. They admit variation as a *vera causa* in one case, they arbitrarily reject it in another, without assigning any distinction in the two cases. The day will come when this will be given as a curious illustration of the blindness of preconceived opinion. These authors seem no more startled at a miraculous act of creation than at an ordinary birth. But do they really believe that at innumerable periods in the earth's history certain elemental atoms have been commanded suddenly to flash into living tissues? Do they believe that at each supposed act of creation one individual or many were produced? Were all the infinitely numerous kinds of animals and plants created as eggs or seed, or as full grown? and in the case of mammals, were they created bearing the false marks of nourishment from the mother's womb? Undoubtedly some of these same questions cannot be answered by those who believe in the appearance or creation of only a few forms of life, or of some one form alone. It has been maintained by several authors that it is as easy to believe in the creation of a million beings as of one; but Maupertuis' philosophical axiom "of least action" leads the mind more willingly to admit the smaller number; and certainly we ought not to believe that innumerable beings within each great class have been created with plain, but deceptive, marks of descent from a single parent.

As a record of a former state of things, I have retained in the foregoing paragraphs, and elsewhere, several sentences which imply that naturalists believe in the separate creation of each species; and I have been much censured for having thus expressed myself. But undoubtedly this was the general belief when the first edition of the present work appeared. I formerly spoke to very many naturalists on the subject of evolution, and never once met with any sympathetic agreement. It is probable that some did then believe in evolution, but they were either silent, or expressed themselves so ambiguously that it was not easy to understand their meaning. Now things are wholly changed, and almost every naturalist admits the great principle of evolution. There are, however, some who still think that species have suddenly given birth, through quite unexplained means, to new and totally different forms: but, as I have attempted to show, weighty evidence can be opposed to the admission of great and abrupt modifications. Under a scientific point of view, and as leading to further investigation, but little advantage is gained by believing that new forms are suddenly developed in an inexplicable manner from old and widely different forms, over the old belief in the creation of species from the dust of the earth.

It may be asked how far I extend the doctrine of the modification of species. The question is difficult to answer, because the more distinct the forms are which we consider, by so much the arguments in favour of community of descent become fewer in number and less in force. But some arguments of the greatest weight extend very far. All the members of whole classes are connected together by a chain of affinities, and all can be classed on the same principle, in groups subordinate to groups. Fossil remains sometimes tend to fill up very wide intervals between existing orders.

Organs in a rudimentary condition plainly show that an early progenitor had the organ in a fully developed condition; and this in some cases implies an enormous amount of modification in the descendants. Throughout whole classes various structures are formed on the same pattern, and at a very early age the embryos closely resemble each other. Therefore I cannot doubt that the theory of descent with modification embraces all the members of the same great class or kingdom. I believe that animals are descended from at most only four or five progenitors, and plants from an equal or lesser number.

Analogy would lead me one step farther, namely, to the belief that all animals and plants are descended from some one proto-type. But analogy may be a deceitful guide. Nevertheless all living things have much in common, in their chemical composition, their cellular structure, their laws of growth, and their liability to injurious influences. We see this even in so trifling a fact as that the same poison often similarly affects plants and animals; or that the poison secreted by the gall-fly produces monstrous growths on the wild rose or oak-tree. With all organic beings excepting perhaps some of the very lowest, sexual production seems to be essentially similar. With all, as far as is at present known the germinal vesicle is the same; so that all organisms start from a common origin. If we look even to the two main divisions—namely, to the animal and vegetable kingdoms—certain low forms are so far intermediate in character that naturalists have disputed to which kingdom they should be referred. As Professor Asa Gray has remarked, "the spores and other reproductive bodies of many of the lower algæ may claim to have first a characteristically animal, and then an unequivocally vegetable existence." Therefore, on the principle of natural selection with divergence of character, it does not seem incredible that, from such low and intermediate form, both animals and plants may have been developed; and, if we admit this, we must likewise admit that all the organic beings which have ever lived on this earth may be descended from some one primordial form. But this inference is chiefly grounded on analogy and it is immaterial whether or not it be accepted. No doubt it is possible, as Mr. G. H. Lewes has urged, that at the first commencement of life many different forms were evolved; but if so we may conclude that only a very few have left modified descendants. For, as I have recently remarked in regard to the members of each great kingdom, such as the Vertebrata, Articulata &c., we have distinct evidence in their embryological homologous and rudimentary structures that within each kingdom all the members are descended from a single progenitor.

When the views advanced by me in this volume, and by Mr. Wallace, or when analogous views on the origin of species are generally admitted, we can dimly foresee that there will be a considerable revolution in natural history. Systematists will be able to pursue their labours as at present; but they will not be incessantly haunted by the shadowy doubt whether this or that form be a true species. This, I feel sure and I speak after experience, will be no slight relief. The endless disputes whether or not some fifty species of British brambles are good species will cease. Systematists will have only to decide (not that this will be easy) whether any form be sufficiently constant and distinct from other forms, to be capable of definition; and if definable, whether the differences be sufficiently important to deserve a specific name. This latter point will become a far more essential consideration than it is at present; for differences, however slight, between any two forms if not blended by intermediate gradations, are looked at by most naturalists as sufficient to raise both forms to the rank of species.

Hereafter we shall be compelled to acknowledge that the only distinction between species and well-marked varieties is, that the latter are known, or believed, to be connected at the present day by intermediate gradations, whereas species were formerly thus connected. Hence, without rejecting the consideration of the present existence of intermediate gradations between any two forms we shall be led to weigh more carefully and to value higher the actual amount of difference between them. It is quite possible that forms now generally acknowledged to be merely varieties may hereafter be thought worthy of specific names; and in this case scientific and common language will come into accordance. In short, we shall have to treat species in the same manner as those naturalists treat genera, who admit that genera are merely artificial combinations made for convenience. This may not be a cheering prospect; but we shall at least be free from the vain search for the undiscovered and undiscoverable essence of the term species.

The other and more general departments of natural history will rise greatly in interest. The terms used by naturalists, of affinity, relationship, community of type, paternity, morphology, adaptive characters, rudimentary and aborted organs, &c., will cease to be metaphorical, and will have a plain signification. When we no longer look at an organic being as a savage looks at a ship, as something wholly beyond his comprehension; when we regard every production of nature as one which has had a long history; when we contemplate every complex structure and instinct as the summing up of many contrivances, each useful to the possessor, in the same way as any great mechanical invention is the summing up of the labour, the experience, the reason, and even the blunders of numerous workmen; when we thus view each organic being, how far more interesting—I speak from experience—does the study of natural history become!

A grand and almost untrodden field of inquiry will be opened, on the causes and laws of variation, on correlation, on the effects of use and disuse, on the direct action of external conditions, and so forth. The study of domestic productions will rise immensely in value. A new variety raised by man will be a more important and interesting subject for study than one more species added to the infinitude of already recorded species. Our classifications will come to be, as far as they can be so made, genealogies; and will then truly give what may be called the plan of creation. The rules for classifying will no doubt become simpler when we have a definite object in view. We possess no pedigrees or armorial bearings; and we have to discover and trace the many diverging lines of descent in our natural genealogies, by characters of any kind which have long been inherited. Rudimentary organs will speak infallibly with respect to the nature of long-lost structures. Species and groups of species which are called aberrant, and which may fancifully be called living fossils, will aid us in forming a picture of the ancient forms of life. Embryology will often reveal to us the structure, in some degree obscured, of the prototype of each great class.

When we feel assured that all the individuals of the same species, and all the closely allied species of most genera, have within a not very remote period descended from one parent, and have migrated from some one birth-place; and when we better know the many means of migration, then, by the light which geology now throws, and will continue to throw, on former changes of climate and of the level of the land, we shall surely be enabled to trace in an admirable manner the former migrations of the inhabitants of the whole world. Even at present, by comparing the differences between the inhabitants of the sea on the opposite sides of a continent, and the nature of the various inhabitants on that continent, in relation to their apparent means of immigration, some light can be thrown on ancient geography.

The noble science of Geology loses glory from the extreme imperfection of the record. The crust of the earth with its imbedded remains must not be looked at as a well-filled museum, but as a poor collection made at hazard and at rare intervals. The accumulation of each great fossiliferous formation will be recognised as having depended on an unusual concurrence of favourable circumstances, and the blank intervals between the successive stages as having been of vast duration. But we shall be able to gauge with some security the duration of these intervals by a comparison of the preceding and succeeding organic forms. We must be cautious in attempting to correlate as strictly contemporaneous two formations, which do not include many identical species, by the general succession of the forms of life. As species are produced and exterminated by slowly acting and still existing causes, and not by miraculous acts of creation; and as the most important of all causes of organic change is one which is almost independent of altered and perhaps suddenly altered physical conditions, namely, the mutual relation of organism to organism,—the improvement of one organism entailing the improvement or the extermination of others; it follows, that the amount of organic change in the fossils of consecutive formations probably serves as a fair measure of the relative though not actual lapse of time. A number of species, however, keeping in a body might remain for a long period unchanged, whilst within the same period several of these species by migrating into new countries and coming into competition with foreign associates, might become modified; so that we must not overrate the accuracy of organic change as a measure of time.

In the future I see open fields for far more important researches. Psychology will be securely based on the foundation already well laid by Mr. Herbert Spencer, that of the necessary acquirement of each mental power and capacity by gradation. Much light will be thrown on the origin of man and his history.

Authors of the highest eminence seem to be fully satisfied with the view that each species has been independently created. To my mind it accords better with what we know of the laws impressed on matter by the Creator, that the production and extinction of the past and present inhabitants of the world should have been due to secondary causes, like those determining the birth and death of the individual. When I view all beings not as special creations, but as the lineal descendants of some few beings which lived long before the first bed of the Cambrian system was deposited, they seem to me to become ennobled. Judging from the past, we may safely infer that not one living species will transmit its unaltered likeness to a distant futurity. And of the species now living very few will transmit progeny of any kind to a far distant futurity; for the manner in which all organic beings are grouped, shows that the greater number of species in each genus, and all the species in many genera, have left no descendants, but have become utterly extinct. We can so far take a prophetic glance into futurity as to foretell that it will be the common and widely-spread species, belonging to the larger and dominant groups within each class, which will ultimately prevail and procreate new and dominant species. As all the living forms of life are the lineal descendants of those which lived long before the Cambrian epoch, we may feel certain that the ordinary succession by generation has never once been broken, and that no cataclysm has desolated the whole world. Hence we may look with some confidence to a secure future of great length. And as natural selection works solely by and for the good of each being, all corporeal and mental endowments will tend to progress towards perfection.

It is interesting to contemplate a tangled bank, clothed with many plants of many kinds, with birds singing on the bushes, with various insects flitting about, and with worms crawling through the damp earth, and to reflect that these elaborately constructed forms, so different from each other, and dependent upon each other in so complex a manner, have all been produced by laws acting around us. These laws, taken in the largest sense, being Growth with Reproduction; Inheritance which is almost implied by reproduction; Variability from the indirect and direct action of the conditions of life, and from use and disuse: a Ratio of Increase so high as to lead to a Struggle for Life, and as a consequence to Natural Selection, entailing Divergence of Character and the Extinction of less-improved forms. Thus, from the war of nature, from famine and death, the most exalted object which we are capable of conceiving, namely, the production of the higher animals, directly follows. There is grandeur in this view of life, with its several powers, having been originally breathed by the Creator into a few forms or into one; and that, whilst this planet has gone cycling on according to the fixed law of gravity, from so simple a beginning endless forms most beautiful and most wonderful have been, and are being evolved.

Selected Readings

The number of books and articles written about Darwin's work is (to use one of Darwin's favorite words) staggering. This brief listing is meant merely to be suggestive. For more comprehensive guidance, one may examine the various bibliographies cited below.

In order not to be cumbersome, works included elsewhere in this volume (except primary Darwin materials) are not repeated here. Titles available in paperback editions are starred.

I. BIBLIOGRAPHICAL MATERIALS

R. B. Freeman, *The Works of Charles Darwin: An Annotated Bibliographical Handlist* (London, 1965).

See also the annual bibliographies in *Isis* and *Victorian Studies* and extensive listings of scientific contributions in Julian Huxley, ★ *Evolution: The Modern Synthesis* (London, 1963) and of sociological contributions in Ashley Montagu, *Darwin: Competition & Cooperation* (New York, 1952). For an extensive discussion of recent studies in various fields, see Bert James Loewenberg, "Darwin and Darwin Studies, 1959–63," *History of Science,* IV (1965), 15–54.

II. BIOGRAPHY

Barlow, Nora, ed., ★ *The Autobiography of Charles Darwin, 1809–1882, with Original Omissions Restored* (New York, 1969).

Barlow, Nora, ed., *Darwin and Henslow: The Growth of an Idea; Letters 1831–1860* (London, 1967).

Darwin, Francis, ed., *The Life and Letters of Charles Darwin* (New York, 1959).

Darwin, Francis and A. C. Seward, eds., *More Letters* (London, 1903).

De Beer, Gavin, ★ *Charles Darwin: A Scientific Biography* (New York, 1964).

III. DARWIN'S WORK

★ *The Structure and Distribution of Coral Reefs* . . . (London, 1842).

Journal of Researches into the Natural History and Geology of the Countries Visited during the Voyage of H. M. S. 'Beagle' . . . (London, 1845).

On the Origin of Species by Means of Natural Selection . . . (London, 1859).

The Variation of Animals and Plants under Domestication (London, 1868).

The Descent of Man . . . (London, 1871).

★ *The Expression of the Emotions in Man and Animals* (London, 1872).

The Formation of Vegetable Mould . . . (London, 1881).

IV. GENERAL OR MISCELLANEOUS STUDIES

Appleman, Philip, ed., ★ *Darwin* (New York, 1970).

Barnett, S. A., ed., *A Century of Darwin* (London, 1958).

Eiseley, Loren C., ★ *Darwin's Century* (New York, 1959).

Ghiselin, Michael T., ★ *The Triumph of the Darwinian Method* (Berkeley and Los Angeles, 1969).

Gruber, Howard and Paul Barrett, *Darwin on Man: A Psychological Study of Scientific Creativity* (New York, 1974).

Huxley, Julian, ★ *Evolution in Action* (New York, 1953).

Huxley, Thomas Henry, *Darwiniana* (New York, 1893).

Persons, Stow, ed., *Evolutionary Thought in America* (New Haven, 1950).

Simpson, George Gaylord, ★ *The Meaning of Evolution* (New Haven, 1949).

Tax, Sol, ed., *Evolution after Darwin* (Chicago, 1960).

Vorzimmer, Peter, *Charles Darwin: The Years of Controversy: The Origin of Species and Its Critics 1859–1882* (Philadelphia, 1970).

V. DARWIN AND SCIENCE

Bateson, W., *Problems of Genetics* (Oxford, 1913).

Bell, P. R., ed., ★ *Darwin's Biological Work* (Cambridge, England, 1959).

De Beer, Gavin R., *Embryos and Ancestors* (Oxford, 1940).

DeVries, Hugo, *Species and Varieties* (Chicago and London, 1905).

Dobzhansky, Theodosius, ★ *Genetics and the Origin of Species* (New York, 1937).

———, ★ *Heredity and the Nature of Man* (New York, 1964).

Fisher, Ronald A., *The Genetical Theory of Natural Selection* (Oxford, 1930).

Gray, Asa, ★ *Darwiniana* (New York, 1876).

Hull, David L., *Darwin and His Critics: The Reception of Darwin's Theory of Evolution by the Scientific Community* (Cambridge, Mass., 1973).

Huxley, Julian, * Evolution: The Modern Synthesis (New York, 1942; revised, 1963).
Mayr, Ernst, Animal Species and Evolution (Cambridge, Mass., 1963).
Morgan, T. H., Evolution and Genetics (Princeton, 1925).
———, The Scientific Basis of Evolution (London, 1932).
Simpson, George Gaylord, Horses (New York, 1961).
Wallace, Alfred Russel, Darwinism (London and New York, 1889).

VI. DARWIN AND PHILOSOPHICAL AND THEOLOGICAL THOUGHT
Bergson, Henri, Creative Evolution (London, 1911).
Dewey, John,* The Influence of Darwin on Philosophy and Other Essays in Contemporary Thought (New York, 1910).
Fiske, John, Darwinism (London and New York, 1879).
* Greene, John C., Darwin and the Modern World View (Baton Rouge, 1961).
* Hardin, Garrett James, Nature and Man's Fate (New York, 1959).
Huxley, Thomas Henry and Julian Huxley, Touchstone for Ethics (New York and London, 1947).
* Medawar, P. B., The Future of Man (London, 1960).
Passmore, John, "Darwin's Impact on British Metaphysics," Victorian Studies, III (1959), 41–54.
Peckham, Morse, "Darwinism and Darwinisticism," Victorian Studies, III (1959), 3–40.
Randall, John Herman, Jr., "The Changing Impact of Darwin on Philosophy," Journal of the History of Ideas, XXII (1961), 435–462.
Raphael, D. D., "Darwinism and Ethics," in Barnett, Darwin's Century (q.v.).
Simpson, George G., * This View of Life (New York, 1964).
* Waddington, C. H., The Ethical Animal (London, 1960).
White, Andrew Dickson,* A History of the Warfare of Science with Theology in Christendom (New York, 1896).

VII. DARWIN AND SOCIETY
Banton, Michael, ed., Darwinism and the Study of Society (London and Chicago, 1961).
Burrow, J. W., Evolution and Society (Cambridge, England, 1966).
Burton, D. H., "Theodore Roosevelt's Social Darwinism and Views on Imperialism," Journal of the History of Ideas, XXVI (1965), 103–118.
Carnegie, Andrew, * The Gospel of Wealth and Other Timely Essays (New York, 1900).
Goldman, Irving, "Evolution and Anthropology," Victorian Studies, III (1959), 55–75.
Gordon, Scott, "Darwinism and Social Thought," in H. H. J. Nesbitt, ed., Darwin in Retrospect (Toronto, 1960).
Hofstadter, Richard, * Social Darwinism in American Thought (Boston, 1955).
Huxley, Julian, Soviet Genetics and World Science (London, 1949).
Kropotkin, Peter, * Mutual Aid (London, 1902).
MacRae, Donald G., "Darwin and the Social Sciences," in Barnett, Darwin's Century (q.v.).
Montagu, Ashley, Darwin: Competition & Cooperation (New York, 1952).
Zirkle, Conway, Evolution, Marxian Biology, and the Social Scene (Philadelphia, 1959).

VIII. DARWIN AND THE LITERARY MIND
* Barzun, Jacques, Darwin, Marx, Wagner: Critique of a Heritage (New York, 1941).
Beach, Joseph Warren, The Concept of Nature in Nineteenth-Century English Poetry (New York, 1936).
* Bush, Douglas, Science and English Poetry (New York, 1950).
* Ellegård, Alvar, Darwin and the General Reader (Göteborg, 1958).
Fleming, Donald, "Charles Darwin, the Anaesthetic Man," Victorian Studies, IV (1961), 219–236.
Henkin, Leo J., Darwinism in the English Novel, 1860–1910 (New York, 1940).
* Hyman, Stanley Edgar, The Tangled Bank. Darwin, Marx, Frazer and Freud as Imaginative Writers (New York, 1962).
Irvine, William, "The Influence of Darwin on Literature," Proceedings of the American Philosophical Society, CIII (1959), 616–628.
Krutch, Joseph Wood, * The Modern Temper (New York, 1929).
* Muller, Herbert J., Science and Criticism (New York, 1956).
* Shaw, George Bernard, Back to Methuselah (London, 1921).
Stevenson, Lionel, Darwin among the Poets (Chicago, 1932).
Willey, Basil, Darwin and Butler: Two Versions of Evolution (London, 1960).

Index